Selenium自动化测试
——基于Python语言

[印度] 昂米沙·冈迪察（Unmesh Gundecha） 著
金鑫　熊志男　译
testwo测试窝　审

人民邮电出版社

北京

图书在版编目（CIP）数据

Selenium自动化测试：基于Python 语言 /（印）冈迪察•U（Unmesh Gundecha）著；金鑫，熊志男译. -- 北京：人民邮电出版社，2018.1（2022.10重印）
ISBN 978-7-115-46174-2

Ⅰ. ①S… Ⅱ. ①冈… ②金… ③熊… Ⅲ. ①软件工具－自动检测 Ⅳ. ①TP311.561

中国版本图书馆CIP数据核字(2017)第274665号

版权声明

Copyright © Packt Publishing 2014. First published in the English language under the title "Learning Selenium Testing Tools with Python", ISBN 978-1-78398-350-6.

All rights reserved.

本书中文简体字版由Packt Publishing 公司授权人民邮电出版社出版。未经出版者书面许可，对本书的任何部分不得以任何方式或任何手段复制和传播。

版权所有，侵权必究。

◆ 著　　[印度] 昂米沙•冈迪察（Unmesh Gundecha）
　译　　金　鑫　熊志男
　审　　Testwo 测试窝
　责任编辑　张　涛
　责任印制　焦志炜

◆ 人民邮电出版社出版发行　北京市丰台区成寿寺路 11 号
邮编 100164　电子邮件 315@ptpress.com.cn
网址 http://www.ptpress.com.cn
北京天宇星印刷厂印刷

◆ 开本：800×1000　1/16
印张：12.75　　　　　　　2018 年 1 月第 1 版
字数：223 千字　　　　　　2022 年 10 月北京第 13 次印刷
著作权合同登记号　图字：01-2016-3957 号

定价：49.00 元
读者服务热线：(010)81055410　印装质量热线：(010)81055316
反盗版热线：(010)81055315
广告经营许可证：京东市监广登字 20170147 号

内容提要

　　Selenium 是一个主要用于 Web 应用程序自动化测试的工具集合，在行业内已经得到广泛的应用。本书介绍了如何用 Python 语言调用 Selenium WebDriver 接口进行自动化测试。主要内容为：基于 Python 的 Selenium WebDriver 入门知识、第一个 Selenium Python 脚本、使用 unittest 编写单元测试、生成 HTML 格式的测试报告、元素定位、Selenium Python API 介绍、元素等待机制、跨浏览器测试、移动端测试、编写一个 iOS 测试脚本、编写一个 Android 测试脚本、Page Object 与数据驱动测试、Selenium WebDriver 的高级特性、第三方工具与框架集成等核心技术。

　　本书适合任何软件测试人员阅读，也适合作为大专院校师生的学习用书和培训学校的教材。

推荐序

认识熊志男是在中国质量大会 BQConf 的活动上,交谈间很快就被志男对于测试领域的见解和趋势展望所折服,而这不仅仅是因为他有丰富的测试经验,而且还因为他作为测试窝的联合创始人,对国内外测试行业有深入的了解。本书正是一个例证。

尽管国内测试行业宣扬自动化测试已经很多年了,但是我们很难在自动化测试领域招聘到经验丰富的工程师,这说明自动化测试并没有成为国内测试领域的主流。

是因为测试人员不够努力吗?我并不这么认为。

很多公司对于测试的投入是希望知道产品有多少缺陷、能否按时上线,所以相应的测试人员的工作都聚焦于如何高效地编写和执行测试用例,而自动化测试并不是第一选择。因为在上线压力巨大的情况下,如果不能评估出自动化测试的投入产出比,很难让项目经理在自动化测试上进行投入。所以倘若有实践自动化测试的想法,往往需要测试人员付出自己的时间来熟悉框架、编写和维护自动化测试。而如果不是事先对测试框架有比较深入的认识,依靠自身自发进行自动化测试,并不会带来效率的明显提升。这样不仅会让项目经理失去信心,恐怕测试人员自己也心存疑虑。

如果单从技术上考虑,究竟是什么阻碍着测试人员广泛使用自动化测试呢?首先,如果没有一个通用的测试框架,那么每做一个项目,测试人员就得学习一套新的工具/框架,这样的学习成本太高了。在工期很紧的情况下更是如此。其次,自动化测试的编写实际上是进行编码,如果使用 Java 和 C#这些编程语言编写自动化测试,测试人员很难全面掌握这些语言的开发技巧,容易导致编写出的自动化测试代码出现比产品代码更多的缺陷。

本书直击这两方面，为测试人员解除了后顾之忧。

（1）Selenium WebDriver 作为业界通用的测试框架，不仅是 Web 测试的标准，在移动测试领域也是底层的核心驱动框架。所以掌握了 Selenium WebDriver，可以让我们在为 Web 产品和移动产品编写自动化测试时游刃有余。

（2）Python 作为动态语言，简化了严格的编程语法，使测试人员更容易掌握。同时 Python 也提供了丰富的 API 和扩展，测试人员可以很便利地调用或者集成其他语言编写的程序和类库，提高编写自动化测试的效率。

本书在讲述自动化测试编写的同时，结合业界主流的自动化测试开发模式，向读者介绍了多种测试相关知识（如 BDD 和持续集成）。非常推荐对测试有激情，希望快速提升自动化测试能力的朋友阅读本书。

黄　勇

现任 ThoughtWorks 中国区 QA Lead

译者序

起初接到本书的翻译邀约时，内心还是有一些困惑的。针对软件测试行业，特别是基于 Web 自动化测试领域，Selenium 已经是广泛使用的工具之一了，而且已被诸多测试同行认可并使用。为此，我们查阅了国内大量相关书籍或文章，发现当前 Selenium 的初学门槛其实不高，测试工程师具备功能测试经验，加之对 Web 前端技术的一定程度的理解，外加较熟练地掌握一门脚本语言，经过一段时间的项目锤炼，都能基本应对日常自动化测试任务。不过与此同时，我们也发现很多初学者遇到的诸多困惑，又或者在深入学习的过程中难以克服的瓶颈。

行业内，能系统性介绍 WebDriver 原理、多类型 Server 运行方式、单元测试以及如何使用 Python 调用 Selenium WebDriver 接口的具体实例的材料相对零散。直到《Learning Selenium Testing Tools with Python》中文版的出版，使得我们有机会较为全面，并且系统地学习用单一脚本语言开发 Web 自动化测试的具体实践，作者独特的创作逻辑，使得本书前后实例相互对照，并且首尾呼应。本书既诠释原理，又能使读者进入实战，还有"干货"满满的"提醒与备注"，是一本不可多得的自动化测试指导书。这也是我们翻译这本书最重要的原因了。

本书的作者 Unmesh Gundecha 有着极为丰富的构建自动化测试解决方案的经验。主导开发过大量商业或开源的自动化测试工具。曾供职微软。在 2012 年编著过《Selenium Testing Tools Cookbook》一书，颇为畅销；在 2015 年下半年又更新发布了第二版。

多年的技术文章翻译经验，使我们清晰地认识到，倘若停留在专业翻译层面，想必本书的可读性，以及作者的诸多表述，都难以顺利地传达给中文读者。所以我与熊志男（本书合译者），多次调整翻译策略。由传统的分章节翻译，到"按作者编著脉络"组织分工。由专业词汇翻译，到统一关键词口径，甚至到整句、整段打乱重组。诸如此类的一些做法，都是为了保证我们的译著质量更加符合测试同行的阅读习惯，便于学习与加深印象。

　　好在翻译过程中，与熊志男相互鼓励，包括审核团队不厌其烦的讨论、PK、一起揣摩作者意图，大家的专业、包容、豁达、自信，让这本书的中文译著得以完成。特别感谢参与翻译工作的张欣欣、谢满彬、谢柳娜。感谢测试窝网译文团队的多次审校。

　　翻译别人的图书，好似在反刍，再精彩也是在讲别人的故事。期待有一天，有机会能够给广大同行讲讲我们自己的故事。由于译者的水平有限，难免会有偏差疏漏。若有欠妥之处，欢迎指正，编辑联系邮箱 zhangtao@ptpress.com.cn。

金　鑫

业界评价

在互联网行业迅速发展的今天，编写自动化脚本的技能，已经逐渐成为 Web 测试人员的标配。

Python 作为备受测试人员青睐的语言之一，非常适合处理日常工作中的数据和文本问题。

Selenium 更是 UI 自动化测试的利器，但要迅速掌握并熟练运用到项目中，绝非易事。

本书围绕 Selenium 的使用展开，编排有序，通俗易懂，对于没有 UI 自动化测试经验的读者，将起到事半功倍的效果。

——Ping++ 质量负责人　吴子腾

Unmesh Gundecha 编著的《*Selenium Testing Tools Cookbook*》，俗称"Selenium 菜谱"，是我一直推荐给身边 WebDriver 初学者的书，只是很遗憾一直未被翻译成中文版本出版。

该书作为"菜谱"的 Python 姊妹篇，秉承了"菜谱"的内容详实、案例丰富、行文流畅等特点，是一本 WebDriver 入门的绝佳教材。

——陈冬严，浙江大学硕士，具有 10 年软件测试和团队管理的工作经验，先后服务于领先的 ITSM、PLM 软件研发企业，现于某金融行业核心机构 IT 规划部门担任项目管理工作。业余时间喜欢园艺。《精通自动化测试框架设计》一书的作者。

关于作者

昂米沙·冈迪察（**Unmesh Gundecha**拥有计算机软件硕士学位，在软件开发与测试领域有着12年的工作经验。无论是在应对业界标准还是定制需求方面，他都有着丰富的构建自动化测试解决方案的经验。与此同时，他还主导开发了大量商业或开源的自动化测试工具。

他曾供职于微软公司，从事与开发有关的工作。目前在印度的一家跨国企业从事测试架构师工作，对于Ruby、Java、iOS、Android和PHP的项目有着极丰富的开发与测试经验。

作者语

另外，本书能顺利编写完成，离不开很多技术同行的帮助与审阅，感谢他们花费了大量的时间为本书提供了非常有价值的反馈。

感谢各位专家、同事和朋友，特别是 Yuri Weinstein 给予我的很多帮助与鼓励。

关于审稿人

Adil Imroz 是一位 Python 的狂热爱好者，长期专注在测试开发与移动端自动化领域。崇尚开源与敏捷模式。闲暇时，爱好骑车、读书、睡觉。他觉得这些都可以为他开拓眼界。

Dr. Philip Polstra（熟悉他的人都称呼他 Dr. Phil），国际知名黑客。他的作品曾在许多国际的专题会议（包括 DEFCON、Black Hat、44CON、Maker Faire 等）上提及，发表过大量的论文，是这一领域公认的专家 Dr. Polstra。

作为布鲁斯伯格大学的副教授，除了日常教学，还对外提供一些关于渗透测试的培训、咨询工作。

Walt Stoneburner，软件架构师，在商业应用开发与咨询领域有着 25 年以上的经验，另外在软件质量保证、配置管理与安全领域也有着长期的研究。

无论是在程序设计、协作应用、大数据、知识管理、数据可视化，还是在 ASCII 方面，他都有着很深的造诣。甚至在软件评测、消费电子产品测评、绘画、经营摄影工作室、创作幽默剧、游戏开发、无线电等领域都能找到他的身影，他还自称"极客"。

Yuri Weinstein，生活在旧金山，有超过 20 年的时间就职于硅谷顶尖的技术公司，专注测试领域，尤其是在自动化测试方向。目前在红帽公司负责管理 Ceph 开源存储项目的产品质量。

前言

Selenium 是一个主要用于 Web 应用程序自动化测试的工具集合，在行业内已经得到广泛的应用。然而它的作用不局限于测试领域，还可以用于屏幕抓取与浏览器行为模拟等操作。它支持主流的浏览器，包括 Firefox、IE、Chrome、Safari 以及 Opera 等。

Selenium 包括一系列的工具组件。

- Selenium IDE——是嵌入 Firefox 浏览器的插件，用于在 Firefox 上录制与回放 Selenium 脚本。图形化的界面可以形象地记录下用户在浏览器中的操作，非常方便使用者了解与学习。目前它只能在 Firefox 下使用，好在它能将录制好的脚本转换成各种 Selenium WebDriver 支持的程序语言，进而扩展到更广泛的浏览器类型。

- Selenium WebDriver——其实质上就是可以支持多种编程语言，并且有用于操作浏览器的一套 API。支持多类型浏览器、跨操作系统平台（包括 Linux、Windows 以及 Mac OS X），是真正意义上的跨浏览器测试工具。WebDriver 为诸如 Java、C#、Python、Ruby、PHP、JavaScript 等语言分别提供了完备的、用于实现 Web 自动化测试的第三方库。

- Selenium Standalone Server——包括被大家广泛了解的 Selenium Grid、远程控制、分布式部署等，均可实现 Selenium 脚本的高效执行与拓展。我们利用 Grid 使得自动化测试可以并行运行，甚至是在跨平台、异构的环境中运行，包括目前主流的移动端环境，如 Android、iOS。

正如书名所述，这是一本介绍如何用 Python 语言调用 Selenium WebDriver 接口，进而实现对 Web 应用自动化测试的指导书。本书描述了从 Selenium 安装配置到基本使用，再到创建、调试、运行自动化脚本等进阶的操作。当然在开始之前，你可能需要先具备一定的 Python 语言基础。

内容介绍

第 1 章基于 Python 的 Selenium WebDriver 入门 从安装 Python、Selenium WebDriver 开始，到我们如何选择适合的 Python 编辑器，以及我们小试牛刀的第一个自动化测试脚本，并且成功地将这一脚本运行在不同浏览器上。

第 2 章使用 unittest 编写单元测试 本章带领我们结合 unittest 实现单元测试。通过转换后的脚本，我们可以更好地完善单元测试用例。借助 unittest 实现测试用例集的整体运行，并将 HTML 格式的测试结果及时推送给项目的相关人员。

第 3 章元素定位 本章告诉你如何通过浏览器自带的开发者模式去定位页面中各类型元素。Selenium 通过获取这些元素的定位，进而实现模拟浏览器操作与参数捕获。这一章你将学会各种定位元素的方法，包括 XPath 和 CSS 以及对应的示例。

第 4 章 Selenium Python API 介绍 学习如何通过 WebDriver 与包括页面元素、JavaScript 提示框、框架（frames）、窗口在内的各类对象进行交互，以及怎样进行浏览器回放、元素传值、鼠标点击、下拉菜单选择、多窗口切换等具体操作。

第 5 章元素等待机制 介绍多种设置等待方法，用于提高 Selenium 自动化测试脚本的稳定运行。带你理解显式等待或隐式等待的方法如何应用于我们的测试脚本。

第 6 章跨浏览器测试 我们将深入学习如何在远程机器或 Selenium Grid 上通过 Remote WebDriver 实现测试脚本跨各类型浏览器的测试。Selenium Grid 可使得我们在多浏览器与多操作系统的排列组合中兼容测试，甚至支持像 PhantomJS 这样的无 UI 界面的浏览器。本章的最后，我们还将了解 Sauce Labs 和 BrowserStack 等第三方外部测试服务（云测试）。

第 7 章移动端测试 我们使用 Selenium WebDriver、Appium 实现在包括 iOS 端、Android 端以及 Android 模拟器在内的移动设备上的自动化测试。另外，本章还有 App 测试的具体示例。

第 8 章 Page Object 与数据驱动测试 介绍这两种重要的设计模式，引导我们搭建更持续、更高效的测试框架。其中，Page Object 设计模式可帮助我们实现对界面细节的封装，并将一组用户行为构建在单个类中，提升自动化测试脚本的易读性和可复用性，从而达到更适应 UI

的频繁变化的目的。另外，我们还将学习用 unittest 实现数据驱动测试。

第 9 章 Selenium WebDriver 的高级特性　包括复杂的鼠标与键盘操作、cookies 操作、屏幕窗口截图，甚至录制整个测试过程。

第 10 章 第三方工具与框架集成　通过 Selenium 与持续集成工具的搭配，我们可以轻松地搭建自动化验收测试框架。本章中展示了"通过 Selenium 创建自动化验收测试用例，然后细化基于 UI 的自动化测试脚本，最后配置持续集成工具 Jenkins，最终实现了对被测程序每日构建、每日自动化验收测试的联动效果"的典型案例。

通过对本书的学习，你将能够用 Python 语言通过调用 Selenium WebDriver 接口，搭建属于你自己的 Web 应用自动化测试框架。

阅读前的准备工作

在阅读本书之前，你需要掌握 Python 语言的基本语法以及 Web 前端的相关知识（如 HTML、JavaScript、CSS 和 XML）。如果你能编写一些简单的包括循环、条件判断、定义类等语法的 Python 脚本，你就能轻松地理解本书中的示例代码。每行示例代码我们都花了很大精力去注释说明，就是希望你能达到最佳的学习效果。还有一些前期准备的软件、工具以及环境配置都在第 1 章有明确的说明，你需要在你的机器上准备好访问终端、Python 解释器以及浏览器。

适合哪些人阅读

如果你从事 QA 或软件测试、软件开发、Web 应用开发等相关工作，希望用 Python 语言调用 Selenium WebDriver，以实现对 Web 应用的自动化测试，那么这本书一定是你较好的选择！在学习 Selenium 理论之前，我们建议你掌握 Python 语言的基本语法。通过整本书的通篇学习，你将全面地理解 Selenium WebDriver 的相关知识，并且能有效地帮助你实现自动化测试。

约定

本书中，你可能会发现不同类型的信息，呈现出的文本风格不尽相同。我们在这里将罗列不同类型的文本风格以及对应的含义，方便你阅读。

代码文本的样式如下。

```
# create a new Firefox session
driver = webdriver.Firefox()
driver.implicitly_wait(30)
driver.maximize_window()
```

当我们想格外强调代码中的一部分时，相应的代码字体会被加粗，示例如下。

```
# run the suite
xmlrunner.XMLTestRunner(verbosity=2,output='test-reports').
 run(smoke_tests)
```

命令行的输入\输出的样式如下。

```
pip install -U selenium
```

新的措辞与**关键语句**会被显示为粗体。关键语句就是指出现在系统界面、菜单项或对话框等位置的关键操作，例如，"在 **Tools** 下拉菜单中选择 **Internet Options**"。

警告或重要的提示，会出现在这样的括号中。

提醒或小窍门，会出现在这样的括号中。

读者反馈

我们十分乐见读者的反馈，让我们了解你对本书的想法——包括好与不好的评价。因为你的反馈将使我们以后可以更好地为读者提供有价值的内容。

可以通过我们的邮箱地址 contact@epubit.com.cn 将你的反馈信息告诉我们，邮件的主题需要注明书籍的全名。

当然，如果你也有专注的主题，并且有兴趣编辑或撰写图书，欢迎你联系我们，邮箱为 zhangtao@ptpress.com.cn。

目录

第 1 章 基于 Python 的 Selenium WebDriver 入门 ········· 1
1.1 环境准备 ········· 2
    1.1.1 安装 Python ········· 3
    1.1.2 安装 Selenium 包 ········· 3
    1.1.3 浏览 Selenium WebDriver Python 文档 ········· 3
    1.1.4 选择一个 IDE ········· 4
    1.1.5 PyCharm 设置 ········· 8
1.2 第一个 Selenium Python 脚本 ········· 11
1.3 支持跨浏览器 ········· 16
    1.3.1 设置 IE 浏览器 ········· 16
    1.3.2 设置 Google Chrome 浏览器 ········· 19
1.4 章节回顾 ········· 21

第 2 章 使用 unittest 编写单元测试 ········· 22
2.1 unittest 单元测试框架 ········· 23
    2.1.1 TestCase 类 ········· 25
    2.1.2 类级别的 setUp()方法和 tearDown()方法 ········· 30
    2.1.3 断言 ········· 32
    2.1.4 测试套件 ········· 33
2.2 生成 HTML 格式的测试报告 ········· 36
2.3 章节回顾 ········· 38

第 3 章 元素定位 ········· 39
3.1 借助浏览器开发者模式定位 ········· 42
    3.1.1 用火狐浏览器 Firebug 插件检查页面元素 ········· 42
    3.1.2 用谷歌 Chrome 浏览器检查页面元素 ········· 43
    3.1.3 用 IE 浏览器检查页面元素 ········· 44
3.2 元素定位 ········· 45
    3.2.1 ID 定位 ········· 46
    3.2.2 name 定位 ········· 47
    3.2.3 class 定位 ········· 47
    3.2.4 tag 定位 ········· 48
    3.2.5 XPath 定位 ········· 50
    3.2.6 CSS 选择器定位 ········· 51
    3.2.7 Link 定位 ········· 53

3.2.8 Partial link 定位 …… 54
3.3 方法实践 …… 54
3.4 章节回顾 …… 58

第4章 Selenium Python API 介绍 …… 59
4.1 HTML 表单元素 …… 60
4.2 WebDriver 原理 …… 61
    4.2.1 WebDriver 功能 …… 61
    4.2.2 WebDriver 方法 …… 61
4.3 WebElement 接口 …… 63
    4.3.1 WebElement 功能 …… 63
    4.3.2 WebElement 方法 …… 63
4.4 操作表单、文本框、复选框、单选按钮 …… 64
    4.4.1 检查元素是否启用或显示 …… 65
    4.4.2 获取元素对应的值 …… 66
    4.4.3 is_selected()方法 …… 67
    4.4.4 clear()与 send_keys()方法 …… 67
4.5 操作下拉菜单 …… 71
    4.5.1 Select 原理 …… 72
    4.5.2 Select 功能 …… 72
    4.5.3 Select 方法 …… 72
4.6 操作警告和弹出框 …… 75
    4.6.1 Alert 原理 …… 75
    4.6.2 Alert 功能 …… 75
    4.6.3 Alert 方法 …… 75
    4.6.4 浏览器自动化处理 …… 78
4.7 章节回顾 …… 80

第5章 元素等待机制 …… 81
5.1 隐式等待 …… 82
5.2 显式等待 …… 84
5.3 expected_conditions 类 …… 85
    5.3.1 判断某个元素是否存在 …… 87
    5.3.2 判断是否存在 Alerts …… 88
5.4 预期条件判断的实践 …… 89
5.5 章节回顾 …… 90

第6章 跨浏览器测试 …… 91
6.1 Selenium Standalone Server …… 93
    6.1.1 下载 Selenium Standalone Server …… 93
    6.1.2 启动 Selenium Standalone Server …… 94
6.2 在 Selenium Standalone Server 上执行测试 …… 95
    6.2.1 配置 IE 支持 …… 98
    6.2.2 配置 Chrome 支持 …… 98
6.3 Selenium Grid …… 98
    6.3.1 启动 hub …… 99
    6.3.2 添加节点 …… 100
6.4 Mac OS X 的 Safari 节点 …… 103
6.5 在 Grid 上执行测试 …… 104
6.6 在云端执行测试 …… 107
6.7 章节回顾 …… 110

第7章 移动端测试 …… 111
7.1 认识 Appium …… 112
    7.1.1 Appium 支持的应用类型 …… 113
    7.1.2 Appium 环境准备 …… 113
7.2 安装 Appium …… 116
7.3 iOS 测试 …… 119
7.4 Android 测试 …… 122
7.5 使用 Sauce Labs …… 126

7.6 章节回顾 ……………………… 128

第 8 章 Page Object 与数据驱动测试 ……………………… 129
8.1 数据驱动测试 ……………………… 130
8.2 使用 ddt 执行数据驱动测试 …… 131
    8.2.1 安装 ddt ……………………… 131
    8.2.2 设计一个简单的数据驱动测试 ……………………… 131
8.3 使用外部数据的数据驱动测试 ……………………… 133
    8.3.1 通过 CSV 获取数据 …… 133
    8.3.2 通过 Excel 获取数据 …… 136
8.4 Page Object 设计模式 …………… 138
    8.4.1 测试准备 ……………………… 140
    8.4.2 BasePage 对象 ……………… 140
    8.4.3 实现 Page Object …………… 141
    8.4.4 构建 Page Object 模式测试实例 ……………………… 145
8.5 章节回顾 ……………………… 146

第 9 章 Selenium WebDriver 的高级特性 ……………………… 147
9.1 键盘与鼠标事件 ……………… 148
    9.1.1 键盘事件 ……………………… 150
    9.1.2 鼠标事件 ……………………… 151
9.2 调用 JavaScript ………………… 154
9.3 屏幕截图 ……………………… 157
9.4 屏幕录制 ……………………… 158
9.5 弹出窗的处理 ………………… 161
9.6 操作 cookies …………………… 163
9.7 章节回顾 ……………………… 165

第 10 章 第三方工具与框架集成 …… 167
10.1 行为驱动开发（BDD） ……… 168
    10.1.1 Behave 安装 ……………… 169
    10.1.2 第一个 feature …………… 169
10.2 持续集成 Jenkins …………… 174
    10.2.1 Jenkins 环境准备 ………… 174
    10.2.2 搭建 Jenkins ……………… 175
10.3 章节回顾 …………………… 182

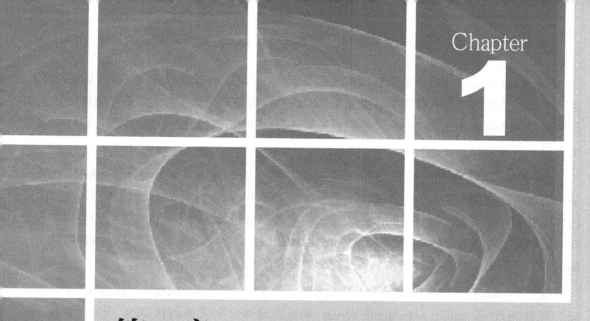

第1章

基于 Python 的 Selenium WebDriver 入门

Selenium 可以自动地操纵浏览器来做很多事情，它可以模拟我们与浏览器的交互，比如，访问网站，单击链接，填写表单，提交表单，浏览网页等，而且支持大多数主流的浏览器。如果要使用 Selenium WebDriver，我们首先要选择一种语言来编写自动化脚本，而这个编程语言需要有 Selenium client library 支持。

本书中，我们将使用支持 Selenium client library 的 Python 语言来编写自动化脚本。Python 是一门被广泛应用的高级编程语言，它是非常容易上手的，而且它的语法使我们只需要简短的代码就可以用来表达思想。Python 设计的初衷就非常强调代码的可读性，它的基础架构使我们可以很方便地写无论大段的还是很少的程序代码，还提供大量的内置库、函数以及用户编写的第三方库，从而能够很容易地实现一些复杂的功能。

基于 Python 的 Selenium WebDriver client library 实现了所有 Selenium WebDriver 特性，而且能够通过 Selenium standalone server 来远程地和分布式地测试 B/S 项目。Selenium language bindings 的开发者包含 David Burns，Adam Goucher，Maik Röder，Jason Huggins，Luke Semerau，Miki Tebeka 和 Eric Allenin。

Selenium WebDriver client library 支持以下 Python 版本：2.6，2.7，3.2 和 3.3。

本章将介绍基于 Python 的 Selenium WebDriver client library 的安装步骤、基本特性和总体架构。

本章包括以下主题：

- 安装 Python 和 Selenium 包；
- 选择和设置 Python 编辑器；
- Selenium WebDriver 基于 Python 编写实例脚本；
- 实现基于 IE 和 Chrome 的跨浏览器支持。

1.1 环境准备

作为学习使用基于 Python 的 Selenium 的第一步，我们需要在计算机上安装好需要的软件。在下面的各节中让我们一步步来配置所需的基础环境。

1.1.1 安装 Python

在安装有 Linux 系统、Mac OS X 系统和其他 UNIX 系统的计算机上，Python 是系统默认安装好的。对于 Windows 系统，就需要另外单独安装 Python 了。基于不同平台的 Python 安装程序都可以在 Python 官方网站找到。

 本书所有的例子都是基于 Python 2.7 和 Python 3.0 编写，并在 Windows 8 系统上经过测试的。

1.1.2 安装 Selenium 包

Selenium 安装包里包含了 Selenium WebDriver Python client library。为了使安装 Selenium 包更简单，可以用 pip 安装工具。

使用 pip，可以非常简单地通过下面的命令来安装和更新 Selenium 安装包。

```
pip install -U selenium
```

安装过程非常简单。该命令将会安装 Selenium WebDriver client library 在计算机上，包含我们使用 Python 来编写自动化脚本需要的所有模块和类。pip 工具将会下载最新版本的 Selenium 安装包并安装在计算机上。这个可选的 -U 参数将会更新已经安装的旧版本至最新版。

也可以从网站下载最新版本的 Selenium 安装包。在页面的右上角单击**下载**按钮，下载后解压文件，然后通过下面的命令来安装。

```
python setup.py install
```

1.1.3 浏览 Selenium WebDriver Python 文档

Selenium WebDriver Python client library 文档可以从 selenium.googlecode 网址查看。

截图如下。

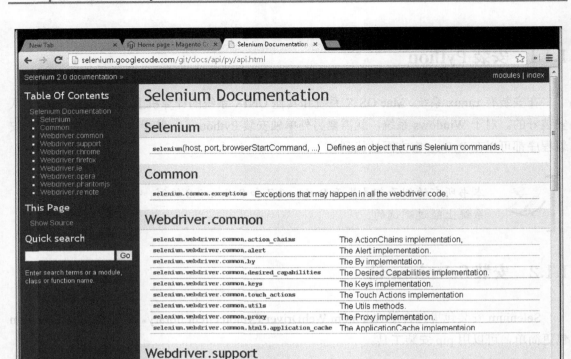

这里提供了 Selenium WebDriver 的所有核心类和函数的详细信息。对于以下链接中的 Selenium 文档也要多加关注。

- 官方文档（seleniumhq 中的 docs）。这里有关于 Selenium 所有组件的说明文档以及基于所支持的语言编写的一些实例。
- Selenium Wiki 地址（google/selenium/w/list）。这里列举了将在后面的章节能够看到的有用的主题。

1.1.4 选择一个 IDE

现在已经安装好了 Python 和 Selenium WebDriver，还需要一个代码编辑器（IDE）来编写自动化脚本。一个好的 **IDE** 能够帮助我们提高产出，而且还能做一些其他的事情让编码变得简单。当我们用简单的编辑器来编写 Python 代码，比如 Emacs、Vim 或 Notepad，用编辑器会让事情变得简单多了。其实有很多的 IDE 可供选择。一般来说，一款好的 IDE 能够通过以下一些特性帮助开发者提高开发速度并节省编码时间：

- 图形化界面编辑器，具备代码编译和代码自动补全功能；
- 方便地查看类和函数代码；
- 语法高亮显示；
- 具备项目管理功能；
- 支持代码模板；
- 具备支持单元测试和调试的工具；
- 支持源码管理。

如果你是个 Python 开发新手，或者是第一次接触 Python 的测试工程师，你的研发同事们能够帮助你安装和配置相应的 IDE。

然而，如果你是第一次接触 Python 而不知道选择哪个 IDE，这里有些建议可以帮助你做出选择。

1.1.4.1 PyCharm

PyCharm 是 JetBrains 公司出品的软件，该公司是专业的软件开发工具的引领者，产品包含大家熟知的 IntelliJ IDEA、RubyMine、PhpStorm 和 TeamCity。

PyCharm 是一款设计精巧、功能强大、应用广泛而且工作良好的 IDE。它继承了 JetBrains 公司其他产品的一贯经验，拥有很多能够提升软件开发效率的特性。

PyCharm 支持 Windows 系统、Linux 系统和 Mac 系统。要想知道更多 PyCharm 的特性，可以访问 jetbrains 网站。

PyCharm 有两种版本——社区版和专业版。社区版是免费的，而专业版是需要付费的。下面是 PyCharm 社区版运行一个简单 Selenium 脚本例子的截图。

社区版能够很好地构建和运行 Selenium 脚本，并提供调试支持。在本书后面的章节将会使用 PyCharm。本章后面的部分，我们一步步来安装 PyCharm，并用它来创建第一个 Selenium 脚本。

本书中所有的例子都是用 PyCharm 构建的，不过读者也可以很容易地应用别的 IDE 来构建这些例子。

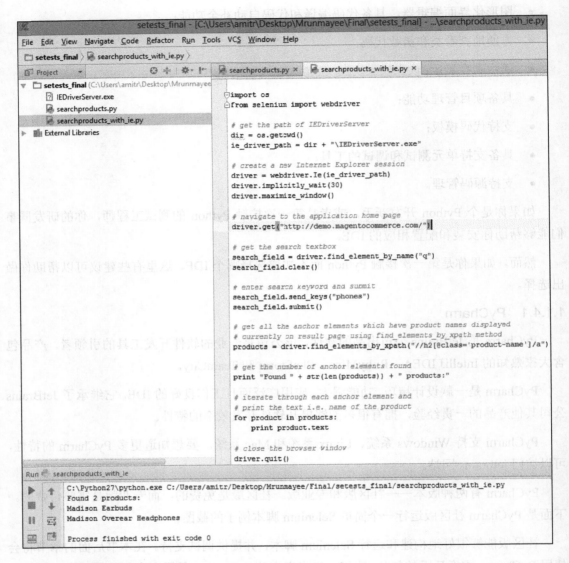

1.1.4.2　PyDev Eclipse plugin

PyDev Eclipse plugin 是另一款在 Python 开发者中应用广泛的代码编辑器。Eclipse 是一款知名的开源代码编辑器，主要应用于构建 Java 程序，它通过插件式架构设计从而实现对其他多种编程语言的支持。

Eclipse 是一款跨平台的 IDE，支持 Windows 系统、Linux 系统和 Mac 系统。可以在 eclipse 网站获取 Eclipse 的最新版本。

PyDev plugin 需要在安装完 Eclipse 后单独安装。可以按照 Lars Vogel 编写的安装指南来安装 PyDev（vogella/tutorials/Python/article.html）。也可以在 pyDev 网站查看安装说明。

下面是使用 PyDev Eclipse plugin 运行一个简单 Selenium 脚本例子的截图。

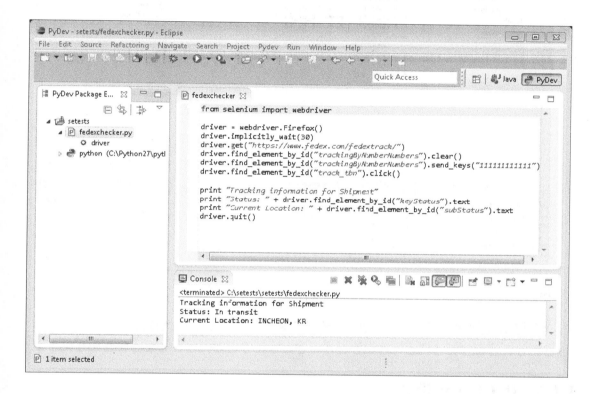

1.1.4.3　PyScripter

对于 Windows 系统的用户，PyScripter 也是一个很好的选择。它是一款开源的、轻量级的 IDE。PyScripter 像其他流行的 IDE 一样具有代码编译和代码自动补全的特性，并且提供测试和调试功能支持。可以从 google 网站下载该软件并获取更多的信息。

下面是在 PyScripter 上运行一个简单 Selenium 脚本实例的截图。

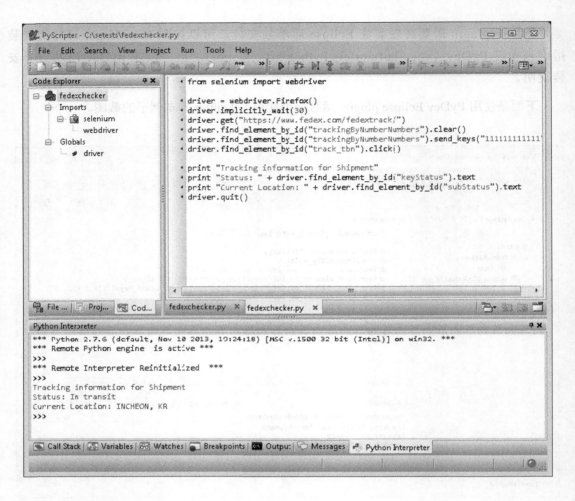

1.1.5 PyCharm 设置

我们看过这些可选择的 IDE 后，回到 PyCharm 的设置。本书中所有的例子脚本都是通过 PyCharm 创建的，不过读者也可以使用其他 IDE 来创建并运行这些实例。下面将通过以下步骤来设置 PyCharm，开始我们的 Selenium Python 之旅。

（1）从 JetBrains 官网下载和安装 PyCharm。

（2）启动 PyCharm 社区版，在启动页面上单击 **Create New Project** 选项。

（3）在 Create New Project 对话框中，参考下面的截图，在 Project name 文本框内输入工程名称。在本例中，setests 作为工程名称。第一次运行 PyCharm 还需要配置解释器。单击 Interpreter 框右侧的 ⋯ 按钮来配置解释器。

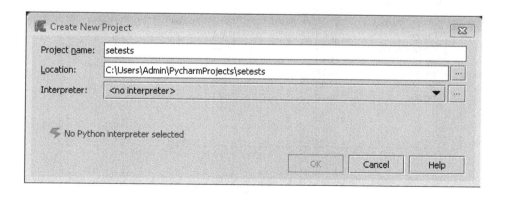

（4）在弹出的 **Python Interpret** 对话框中，单击加号，PyCharm 会显示出已经安装好的解释器路径。从 Select Interpreter Path 中选择对应的解释器。

（5）PyCharm 将会配置好刚才选择的解释器。在 Packages 选项卡会显示 Python 安装包内自带的一些工具包。单击 **Apply** 按钮，然后单击 **OK** 按钮。

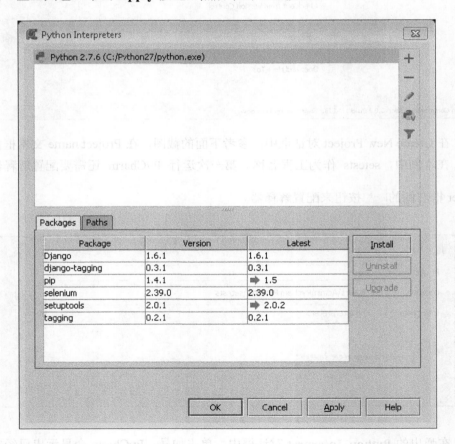

（6）返回到 **Create New Project** 对话框，单击 **OK** 按钮，项目创建成功。

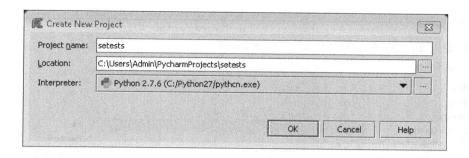

1.2 第一个 Selenium Python 脚本

我们现在可以开始创建和运行自动化测试脚本了。就从 Selenium WebDriver 开始，然后创建一个 Python 脚本，用 Selenium WebDriver 提供的类和方法模拟用户与浏览器的交互。

我们会使用一个简单的 Web 应用程序（本书上大多数例子都是基于这个应用程序）。这个简单的 Web 应用程序是基于一个著名电子商务框架 **Magento** 构建的。你可以在 magentocommerce 网站找到这个应用程序。

示例代码下载
如果你是在 packtpub 购买本书，你可以通过你的账号在该网址上下载示例代码文件。如果你是从其他地方购买本书，你可以访问 packtpub 网站并注册，我们会把示例代码文件直接发送到你的邮箱中。
示例代码也被托管在 github 中，访问地址为：github/upgundecha/learnsewithpython。

在这个简单的脚本中，我们会通过接下来的步骤去访问这个应用程序，搜索产品并在搜索结果页面中列出产品的名称。

（1）我们使用早前在部署 PyCharm 环境时创建的项目。创建一个引用了 Selenium WebDriver client library 的 Python 脚本。在项目资源管理器视图中，用鼠标右键单击 `setests`，在弹出的菜单中依次选择 **New→Python File**。

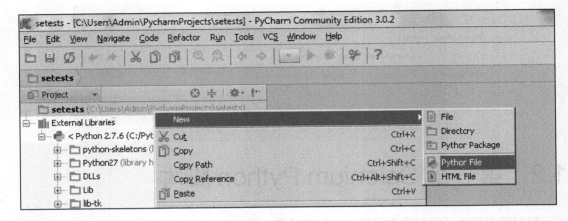

（2）在 **New Python file** 对话框中，在 Name 文本框中输入"searchproducts"，然后单击 OK 按钮。

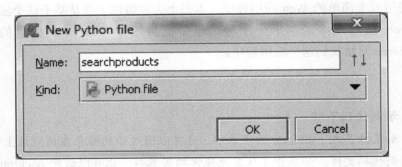

（3）PyCharm 会在代码编写区域增加一个名为 searchproducts.py 的新页签。复制下列代码到 searchproducts.py 中。

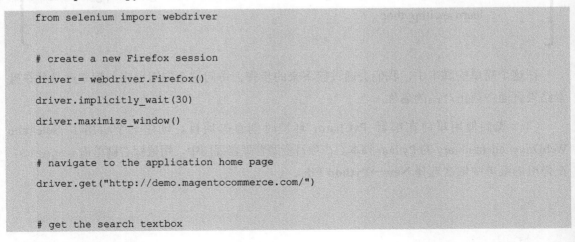

```
search_field = driver.find_element_by_name("q")
search_field.clear()

# enter search keyword and submit
search_field.send_keys("phones")
search_field.submit()

# get all the anchor elements which have product names displayed
# currently on result page using find_elements_by_xpath method
products = driver.find_elements_by_xpath("//h2[@class='product-name']/a")

# get the number of anchor elements found
print "Found " + str(len(products)) + " products:"

# iterate through each anchor element and print the text that is
# name of the product
for product in products:
    print product.text

# close the browser window
driver.quit()
```

 如果你使用的是其他 IDE 编译工具，请同样创建一个新的文件，复制代码到文件中并保存为 searchproducts.py。

（4）可以通过以下方式运行脚本：在 PyCharm 代码窗口中使用快捷键 Ctrl + Shift+ F10 或者在 **Run** 菜单中选择 **Run 'searchproducts'** 命令。脚本开始执行，你会看到新弹出一个 Firefox 浏览器窗口访问演示网址，接着在 Firefox 浏览器窗口中会看到被执行的 Selenium 命令。如果一切运行顺利，最后脚本会关闭 Firefox 浏览器窗口。如下图所示，这个脚本会在 PyCharm 的控制台中打印产品的清单。

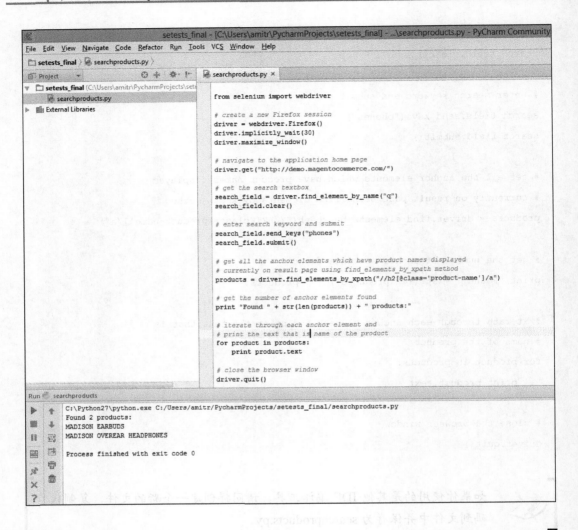

我们也可以在命令行中运行脚本。打开命令行工具，切换到 setests 项目所在的目录中，运行以下命令：

```
python searchproducts.py
```

在本书后面部分，我们更喜欢选择使用命令行方式去执行测试脚本。

接下来，我们将会花点时间分析刚才创建的脚本。我们分析每个语句，初步地认识 Selenium WebDriver。在本书的后面部分还有很多这样的分析。

Selenium.webdriver 模块实现了 Selenium 所支持的各种浏览器驱动程序类，包括 Firefox 浏览器、Chrome 浏览器、IE 浏览器、Safari 浏览器和多种其他浏览器。另外，RemoteWebDriver

则是用于调用远程机器进行浏览器测试的。

我们需要从 Selenium 包中导入 WebDriver 才能使用 Selenium WebDriver 方法。

```
from selenium import webdriver
```

接着，我们还需要选用一个浏览器驱动实例，它会提供一个接口去调用 Selenium 命令来跟浏览器交互。在这个例子中，我们使用的是 Firefox 浏览器。我们可以通过下方命令来创建一个 Firefox 浏览器驱动实例。

```
driver = webdriver.Firefox()
```

在运行期间，这会加载一个新的 Firefox 浏览器窗口。我们也可以在这个驱动上设置一些参数，如：

```
driver.implicityly_wait(30)
driver.maximize_window()
```

我们使用 30 秒隐式等待时间来定义 Selenium 执行步骤的超时时间，并且调用 Selenium API 来最大化浏览器窗口。我们会在第 5 章"元素等待机制"中学习更多关于隐式等待的内容。

接着，我们使用示例程序的 URL 作为参数，通过调用 driver.get() 方法访问该应用程序。在 get() 方法被调用后，WebDriver 会等待，一直到页面加载完成才继续控制脚本。

在加载页面后，Selenium 会像用户真实使用那样，和页面上各种各样的元素交互。例如，在应用程序的主页，我们需要在输入框中输入一个搜索内容，然后单击 **Search** 按钮。这些元素作为 HTML 输入元素实现，Selenium 需要找到这些元素来模拟用户操作。Selenium WebDriver 提供多种方法来定位和操作这些元素，例如设置值，单击按钮，在下拉组件中选择选项等。我们可以在第 3 章"元素定位"中了解更多。

在这个例子中，我们使用 find_element_by_name 方法来定位搜索输入框。这个方法会返回第一个 name 属性值与输入参数匹配的元素。HTML 元素是用标签和属性来定义的，我们可以使用这些信息来定位一个元素，步骤如下。

（1）在这个例子中，搜索输入框有一个值为 q 的 name 属性，我们使用这个属性来定位，代码如下。

```
search_field = driver.find_element_by_name("q")
```

（2）一旦找到这个搜索输入框，我们可以使用 clear() 方法来清理之前的值（如果搜索输入框已经有值的话），并且通过 send_keys() 方法输入新的特定的值。接着我们通过调用 submit() 方法提交搜索请求。

```
search_field.clear()
search_field.send_keys("phones")
search_field.submit()
```

（3）在提交搜索请求后，Firefox 浏览器会加载结果页面。结果页面中有一系列与搜索项（phones）匹配的产品。我们可以读取结果列表，并且可以使用 find_elements_by_xpath 方法获取路径是以<a>标签结尾的所有产品名称。它将会返回多于 1 个的元素列表。

```
products = 
  driver.find_elements_by_xpath("//h2[@class=
  'product-name']/a")
```

（4）接着，我们打印在页面中展示的产品个数（即符合路径以<a>标签结尾的元素个数）和产品的名称（即<a>标签的 text 属性值）。

```
print "Found " + str(len(products)) + " products:"

for product in products:
    print product.text
```

（5）在脚本的最后，我们使用 driver.quit()方法来关闭 Firefox 浏览器。

```
driver.quit()
```

这个例子直观地向我们展示了如何使用 Selenium WebDriver 和 Python 配合来创建一个简单的自动化脚本。我们在这个脚本里面并没有测试什么。在本书后面的章节，我们将会扩展这个简单的脚本为一组测试脚本，并且会引用多个其他库和 Python 的功能。

1.3 支持跨浏览器

目前我们已经在 Firefox 浏览器构建并运行了脚本。Selenium 支持各种浏览器，读者可以在不同的浏览器中进行自动化测试。它支持的浏览器包括 IE 浏览器、Google Chrome 浏览器、Safari 浏览器、Opera 浏览器，甚至是像 PhantomJS 这样的无 UI 界面的浏览器。接下来的部分，我们会修改刚才创建的脚本，以便在 IE 浏览器和 Google Chrome 浏览器中运行脚本，以此来验证 Selenium WebDriver 跨浏览器的兼容性。

1.3.1 设置 IE 浏览器

在 IE 浏览器中运行脚本的步骤会多一些。我们需要下载并安装 InternetExplorerDriver。

InternetExplorerDriver 是一个独立的可执行的服务，它实现 WebDriver 的协议，使得 WebDriver 可以与测试脚本和 IE 浏览器交互。InternetExplorerDriver 支持 Windows XP、Vista、Windows 7 和 Windows 8 操作系统下的主要 IE 版本。通过以下步骤安装 InternetExplorerDriver。

（1）在 seleniumhq 网站中下载 InternetExplorerDriver 服务。你可以根据自己的操作系统来选择下载 32 位或 64 位的版本。

（2）在下载完成后，解压文件，并把文件复制到存储脚本的目录中。

（3）在 IE 7 及其以上版本，每个区域的**保护模式**设置一定要有相同的值。在每个区域中**保护模式**要么启用，要么关闭。设置**保护模式**的步骤如下。

① 在**工具**菜单下选择 Internet 选项。

② 在 **Internet 选项**对话框中，单击**安全**标签页。

③ 在"**选择区域以查看或更改安全设置**"中选择每一个区域，确定每个区域的保护模式的值保持一致（要么选中，要么不选中）。所有区域用相同的设置，如下图所示。

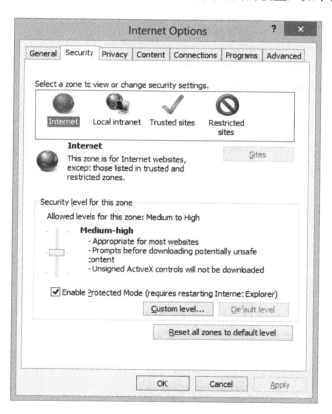

> 在使用 InternetExplorerDriver 时,注意保持浏览器缩放等级设置成 100%,以此来保证鼠标的单击事件能点到正确的坐标。

(4)修改脚本使其支持 IE 浏览器。我们通过以下方式来使用 IE 替代 Firefox 实例。

```python
import os
from selenium import webdriver

# get the path of IEDriverServer
dir = os.path.dirname(__file__)
ie_driver_path = dir + "\IEDriverServer.exe"

# create a new Internet Explorer session
driver = webdriver.Ie(ie_driver_path)
driver.implicitly_wait(30)
driver.maximize_window()

# navigate to the application home page
driver.get("http://demo.magentocommerce.com/")

# get the search textbox
search_field = driver.find_element_by_name("q")
search_field.clear()

# enter search keyword and submit
search_field.send_keys("phones")
search_field.submit()

# get all the anchor elements which have product names displayed
# currently on result page using find_elements_by_xpath method
products = driver.find_elements_by_xpath("//h2[@class='product-name']/a")

# get the number of anchor elements found
print "Found " + str(len(products)) + " products:"
```

```
# iterate through each anchor element and print the text that is
# name of the product
for product in products:
    print product.text


# close the browser window
driver.quit()
```

在这个脚本中，在创建 IE 浏览器实例时，我们传递了 InternetExplorerDriver 的路径。

（5）运行脚本后，Selenium 会加载 InternetExplorerDriver 服务，用它来启动浏览器和执行脚本。InternetExplorerDriver 服务在 Selenium 脚本和浏览器之间扮演类似中介角色。实际执行的步骤与我们在 Firefox 浏览器观察的类似。

在 code.google/p/selenium/wiki/
InternetExplorerDriver 中可以获取更多关于 IE 的重要设置。
在 code.google/p/selenium/wiki/
DesiredCapabilities 查阅 DesiredCapabilities 的文章。

1.3.2 设置 Google Chrome 浏览器

在 Google Chrome 浏览器中设置和运行脚本的步骤与 IE 浏览器的相似。我们需要下载 ChromeDriver 服务。ChromeDriver 服务是一个由 Chromium team 开发维护的独立的服务，它支持 Windows 操作系统、Linux 操作系统和 Mac 操作系统。使用以下步骤来设置 ChromeDriver 服务。

（1）在 chromedriver.storage.googleapis/index.html 下载 ChromeDriver 服务。

（2）下载完 ChromeDriver 服务后，解压文件，并把文件复制到存储脚本的目录中。

（3）修改脚本使其支持 Chrome 浏览器。我们通过以下方式创建 Chrome 实例，用此来替换 Firefox 浏览器实例。

```
import os
from selenium import webdriver
```

```python
# get the path of chromedriver
dir = os.path.dirname(__file__)
chrome_driver_path = dir + "\chromedriver.exe"
# remove the .exe extension on linux or mac platform

# create a new Chrome session
driver = webdriver.Chrome(chrome_driver_path)
driver.implicitly_wait(30)
driver.maximize_window()

# navigate to the application home page
driver.get("http://demo.magentocommerce.com/")

# get the search textbox
search_field = driver.find_element_by_name("q")
search_field.clear()

# enter search keyword and submit
search_field.send_keys("phones")
search_field.submit()

# get all the anchor elements which have product names displayed
# currently on result page using find_elements_by_xpath method
products = driver.find_elements_by_xpath("//h2[@class='product-name']/a")

# get the number of anchor elements found
print "Found " + str(len(products)) + " products: "

# iterate through each anchor element and print the text that is
# name of the product
for product in products:
    print product.text

# close the browser window
```

```
driver.quit()
```

在这个脚本中，在创建 Chrome 浏览器实例时，我们传递了 ChromeDriver 的路径。

（4）运行脚本后，Selenium 会加载 ChromeDriver 服务，用它来启动浏览器和执行脚本。实际执行的步骤与我们在 Firefox 浏览器观察的类似。

想了解更多关于 ChromeDriver，请访问 code.google/p/selenium/ wiki/ChromeDriver 和 sites.google/a/chromium.org/chromedriver/ home。

1.4 本章回顾

在本章中，我们介绍了 Selenium 和它的组件。通过 pip 工具安装了 Selenium 包。接着介绍了多个用于编写 Selenium 和 Python 代码的编辑器和 IDE 工具，部署了 PyCharm 环境。然后我们通过创建基于示例程序的测试脚本，成功运行在 Firefox 浏览器，并分析了整个过程。最后，我们举一反三（分别在 IE 浏览器和 Chrome 浏览器中配置和运行脚本）来验证 Selenium WebDriver 的跨浏览器的特性。

在下一章，我们将学习如何通过 Selenium WebDriver 使用 unittest 库来创建自动化单元测试。我们也将学习如何创建并运行一组测试脚本。

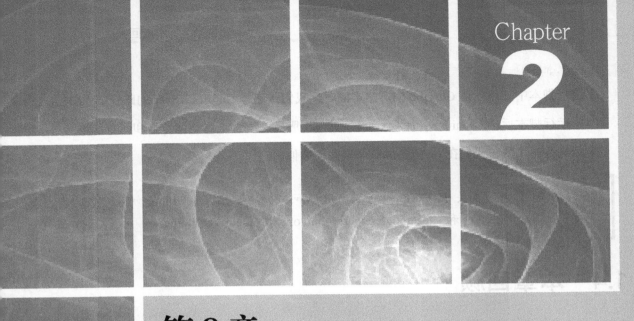

第 2 章

使用 unittest 编写单元测试

Selenium WebDriver 是一个浏览器自动化测试的 API 集合。它提供了很多与浏览器自动化交互的特性，并且这些 API 主要是用于测试 Web 程序。如果仅仅使用 Selenium WebDriver，我们无法实现执行测试前置条件、测试后置条件，比对预期结果和实际结果，检查程序的状态，生成测试报告，创建数据驱动的测试等功能。在本章，我们将学习如何使用 unittest 来创建基于 Python 的 Selenium WebDriver 测试脚本。

本章包含以下主题：

- 什么是 unittest？
- 使用 unittest 来写 Selenium WebDriver 测试；
- 用 TestCase 类来实现一个测试；
- 学习 unittest 提供的不同类型的 assert 方法；
- 为一组测试创建 TestSuite；
- 使用 unittest 扩展来生成 HTML 格式的测试报告。

2.1　unittest 单元测试框架

我们可以使用 unittest 为任何项目创建全面的测试套件。unittest 也是 Python 中用来测试各种标准类库模块的。可以在 Python 网站查看 unittest 的文档。

unittest 使我们具备创建测试用例、测试套件、测试夹具的能力。可以通过下面的图来了解所有的组件。

- **Test Fixture**（测试夹具）：通过使用测试夹具，可以定义在单个或多个测试执行之前的准备工作和测试执行之后的清理工作。
- **Test Case**（测试用例）：一个测试用例是在 unittest 中执行测试的最小单元。它通过 unittest 提供的 assert 方法来验证一组特定的操作和输入以后得到的具体响应。unittest 提供了一个名称为 TestCase 的基础类，可以用该类来创建测试用例。
- **Test Suite**（测试套件）：一个测试套件是多个测试或测试用例的集合，是针对被测程序的对应的功能和模块创建的一组测试，一个测试套件内的测试用例将一起执行。

- **Test Runner**（测试执行器）：测试执行器负责测试执行调度并且生成测试结果给用户。测试执行器可以使用图形界面、文本界面或者特定的返回值来展示测试执行结果。
- **Test Report**（测试报告）：测试报告用来展示所有执行用例的成功或者失败状态的汇总，执行失败的测试步骤的预期结果与实际结果，还有整体运行状况和运行时间的汇总。

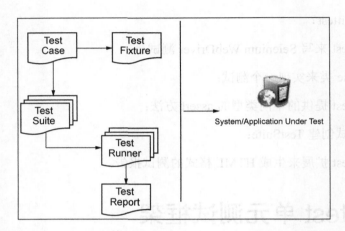

通过与 unittest 类似的 xUnite 测试框架创建的测试被拆分为 3 部分，即 3A's，具体如下。

- **Arrange**：用来初始化测试的前置条件，包含初始化被测试的对象、相关配置和依赖。
- **Act**：用来执行功能操作。
- **Assert**：用来校验实际结果与预期结果是否一致。

我们在本章接下来的内容中将应用此方法来使用 unittest 创建测试。

我们将在本书接下来的部分使用 unittest 来创建和运行基于 Selenium WebDriver 的测试。另外，Python 还有些具备额外特性的其他测试框架，例如：

- **Nose**：此框架扩展了 unittest 并且提供了自动搜索和运行测试的功能，也提供了一些插件来创建高级的测试。可以在 readthedocs 网站查看关于 Nose 的更多信息。
- **Pytest**：Pytest 是另外一个测试框架，它提供了一些基于 Python 来编写和运行单元测试的高级特性。可以在 pytest 网站查看关于 Pytest 的更多信息。

2.1.1 TestCase 类

我们可以通过继承 TestCase 类并且在测试类中为每一个测试添加测试方法来创建单个测试或者一组测试。为了创建测试，我们需要使用 TestCase 类中的 assert 或者使用其中的一种 assert 方法。每个测试最重要的任务是调用 assertEqual() 来校验预期结果，调用 assertTrue() 来验证条件，或者调用 assertRaises() 来验证预期的异常。

除了添加测试，我们可以添加测试夹具——setUp()方法和 tearDown()方法，创建或处置测试用例所需要的任何对象和条件。

让我们开始使用 unittest，首先通过继承 TestCase 类然后，添加一个测试方法，来为第 1 章（基于 Python 的 Selenium WebDriver 入门）中的例子脚本写一个简单的测试。

我们需要先引入 unittest 模块，然后定义一个继承于 TestCase 类的子类，具体如下。

```
import unittest
from selenium import webdriver


class SearchTest (unittest.TestCase):
```

2.1.1.1 setUp()方法

一个测试用例是从 setUp()方法开始执行的，我们可以用这个方法在每个测试开始前去执行一些初始化的任务。可以是这样的初始化准备：比如创建浏览器实例，访问 URL，加载测试数据和打开日志文件等。

此方法没有参数，而且不返回任何值。当定义了一个 setUp()方法，测试执行器在每次执行测试方法之前优先执行该方法。在下面的例子里，我们将用 setUp()方法来创建 Firefox 的实例，设置 properties，而且在测试开始执行之前访问被测程序的主页。例子如下。

```python
import unittest
from selenium import webdriver


class SearchTests(unittest.TestCase):
    def setUp(self):
        # create a new Firefox session
        self.driver = webdriver.Firefox()
        self.driver.implicitly_wait(30)
        self.driver.maximize_window()

        # navigate to the application home page
        self.driver.get("http://demo.magentocommerce.com/")
```

2.1.1.2 编写测试

有了 setUp()方法，现在可以写一些测试来验证我们想要测试的程序的功能。在这个例子里，我们将搜索一个产品，然后检查是否返回一些相应的结果。与 setUp()方法相似，test 方法也是在 TestCase 类中实现。重要的一点是我们需要给测试方法命名为 test 开头。这种命名约定通知 test runner 哪个方法代表测试方法。

对于 test runner 能找到的每个测试方法，都会在执行测试方法之前先执行 setUp()方法。这样做有助于确保每个测试方法都能够依赖相同的环境，无论类中有多少测试方法。我们将使用简单的 assertEqual()方法来验证用程序搜索该术语返回的结果是否和预期结果相匹配。我们将在本章后面内容中探讨更多关于断言的内容。

添加一个新的测试方法 test_search_by_category()，通过分类来搜索产品，然后校验返回的产品的数量是否正确，具体如下。

```python
import unittest
from selenium import webdriver


class SearchTests(unittest.TestCase):
    def setUp(self):
```

```
        # create a new Firefox session
        self.driver = webdriver.Firefox()
        self.driver.implicitly_wait(30)
        self.driver.maximize_window()

        # navigate to the application home page
        self.driver.get("http://demo.magentocommerce.com/")

    def test_search_by_category(self):
        # get the search textbox
        self.search_field = self.driver.find_element_by_name("q")
        self.search_field.clear()

        # enter search keyword and submit
        self.search_field.send_keys("phones")
        self.search_field.submit()

        # get all the anchor elements which have product names
        # displayed currently on result page using
        # find_elements_by_xpath method
        products = self.driver.find_elements_by_xpath
          ("//h2[@class='product-name']/a")
        self.assertEqual(2, len(products))
```

2.1.1.3 代码清理

类似于 setUp() 方法在每个测试方法之前被调用，TestCase 类也会在测试执行完成之后调用 tearDown() 方法来清理所有的初始化值。一旦测试被执行，在 setUp() 方法中定义的值将不再需要，所以最好的做法是在测试执行完成的时候清理掉由 setUp() 方法初始化的数值。在我们的例子里，在测试执行完成后，就不再需要 Firefox 的实例。我们将在 tearDown() 方法中关闭 Firefox 实例，如下代码所示。

```
import unittest
from selenium import webdriver


class SearchTests(unittest.TestCase):
```

```python
    def setUp(self):
        # create a new Firefox session
        self.driver = webdriver.Firefox()
        self.driver.implicitly_wait(30)
        self.driver.maximize_window()

        # navigate to the application home page
        self.driver.get("http://demo.magentocommerce.com/")

    def test_search_by_category(self):
        # get the search textbox
        self.search_field = self.driver.find_element_by_name("q")
        self.search_field.clear()

        # enter search keyword and submit
        self.search_field.send_keys("phones")
        self.search_field.submit()

        # get all the anchor elements which have product names
        # displayed currently on result page using
        # find_elements_by_xpath method
        products = self.driver.find_elements_by_xpath
          ("//h2[@class='product-name']/a")
        self.assertEqual(2, len(products))

    def tearDown(self):
        # close the browser window
        self.driver.quit()
```

2.1.1.4 运行测试

为了通过命令行运行测试，我们可以在测试用例中添加对 main 方法的调用。我们将传递 verbosity 参数以便使详细的测试总量展示在控制台。

```python
if __name__ == '__main__':
    unittest.main(verbosity=2)
```

我们可以把测试脚本保存为普通的 Python 脚本。在这个例子里，把测试保存为 searchtests.py。保存文件以后，我们可以通过下面的命令行来执行该测试。

```
python searchtests.py
```

测试运行结束后，unittest 会把测试结果和概要展示在控制台，如下图所示。

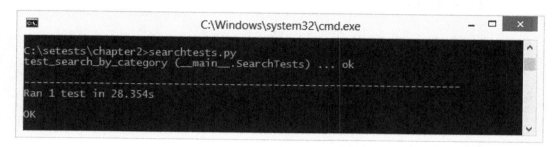

除了测试结果概要外，当一个测试用例执行失败，针对每个失败，测试结果概要都会通过生成文本信息来展示具体哪里有错误。通过下面的截图，可以看到当我们修改预期结果值后会发生些什么。

上图展示了具体是哪个测试方法执行失败,通过打印信息可以追踪具体导致失败的代码。另外，失败自身也会以 AssertionError 形式显示，例子中的预期结果和实际结果并不匹配。

2.1.1.5 添加其他测试

我们可以用一组测试来构建一个测试类，这样有助于为一个特定功能创建一组更合乎逻辑的测试。下面为测试类添加其他的测试。规则很简单，新的测试方法命名也要以 test 开头，如下列代码。

```python
def test_search_by_name(self):
    # get the search textbox
    self.search_field = self.driver.find_element_by_name("q")
    self.search_field.clear()

    # enter search keyword and submit
    self.search_field.send_keys("salt shaker")
    self.search_field.submit()

    # get all the anchor elements which have
    # product names displayed
    # currently on result page using
    # find_elements_by_xpath method
    products = self.driver.find_elements_by_xpath
       ("//h2[@class='product-name']/a")
    self.assertEqual(1, len(products))
```

运行这个测试类将能看到两个 Firefox 的实例打开和关闭，这正是 setUp()方法和 tearDown()方法针对每个测试方法都要执行产生的结果，如下图所示。

```
C:\setests\chapter2>searchtests.py
test_search_by_category (__main__.SearchTests) ... ok
test_search_by_name (__main__.SearchTests) ... ok

----------------------------------------------------------------------
Ran 2 tests in 61.746s

OK
```

2.1.2 类级别的 setUp()方法和 tearDown()方法

在前面的例子中，我们通过 setUp()方法为每个测试方法都创建了一个 Firefox 实例，并且在每个测试方法执行结束后都要关闭实例。能否让各个测试方法共用一个 Firefox 实例，而不要每次都创建一个新的实例呢？这可以通过使用 setUpClass()方法和 tearDownClass()方法及@classmethod 标识来实现。这两个方法使我们可以在类级别来初始化数据，替代了方法级别的初始化，这样各个测试方法就可以共享这些初始化数据。在下面的例子中，代码修改为调用 setUpClass()方法和 tearDownClass()方法并且加上@classmethod 标识。

```python
import unittest
from selenium import webdriver

class SearchTests(unittest.TestCase):
    @classmethod
    def setUpClass(cls):
        # create a new Firefox session
        cls.driver = webdriver.Firefox()
        cls.driver.implicitly_wait(30)
        cls.driver.maximize_window()

        # navigate to the application home page
        cls.driver.get("http://demo.magentocommerce.com/")
        cls.driver.title

    def test_search_by_category(self):
        # get the search textbox
        self.search_field = self.driver.find_element_by_name("q")
        self.search_field.clear()

        # enter search keyword and submit
        self.search_field.send_keys("phones")
        self.search_field.submit()

        # get all the anchor elements which have product names
        # displayed currently on result page using
        # find_elements_by_xpath method
        products = self.driver.find_elements_by_xpath
            ("//h2[@class='product-name']/a")
        self.assertEqual(2, len(products))

    def test_search_by_name(self):
```

```python
    # get the search textbox
    self.search_field = self.driver.find_element_by_name("q")
    self.search_field.clear()

    # enter search keyword and submit
    self.search_field.send_keys("salt shaker")
    self.search_field.submit()

    # get all the anchor elements which have product names
    # displayed currently on result page using
    # find_elements_by_xpath method
    products = self.driver.find_elements_by_xpath
      ("//h2[@class='product-name']/a")
    self.assertEqual(1, len(products))

@classmethod
def tearDownClass(cls):
    # close the browser window
    cls.driver.quit()

if __name__ == '__main__':
    unittest.main()
```

运行这个测试将看到仅创建一个 Firefox 实例,所有的测试都用同一个实例。

要了解更多关于@classmethod 标识的信息,参考:
docs.python/2/library/ functions.html#classmethod。

2.1.3 断言

unittest 的 TestCase 类提供了很多实用的方法来校验预期结果和程序返回的实际结果是否一致。这些方法要求必须满足某些条件才能继续执行接下来的测试。大致有 3 种这样的方法,各覆盖一个特定类型的条件,例如等价校验、逻辑校验和异常校验。如果给定的断言通过了,接下来的测试代码将会执行;相反,将会导致测试立即停止并且给出异常信息。

unittest 提供了所有的标准 xUnit 断言方法。下表列出了一些在本书后面将要用到的重要方法。

方　　法	校 验 条 件	应 用 实 例
assertEqual(a, b [,msg])	a == b	这些方法校验 a 和 b 是否相等，msg 对象是用来说明失败原因的消息。 这对于验证元素的值和属性等是非常有用的。 例如： assertEqual(element.text, "10")
assertNotEqual(a, b[,msg])	a != b	
assertTrue(x[,msg]))	bool(x) is True	这些方法校验给出的表达式是 True 还是 False。 例如，校验一个元素是否出现在页面，我们可以用下面的方法： assertTrue(element.is_displayed())
assertFalse(x[,msg]))	bool(x) is False	
assertIsNot(a, b[,msg]))	a is not b	
assertRaises(exc, fun, *args, **kwds)	fun(*args, **kwds) raises exc	这些方法校验特定的异常是否被具体的测试步骤抛出，用到该方法的一种可能情况是： NoSuchElementFoundexception
assertRaisesRegexp(exc, r, fun, *args, **kwds)	fun(*args, **kwds) raises exc and the message matches regex r	
assertAlmostEqual(a, b)	round(a-b, 7) == 0	这些方法用于检查数值，在检查之前会按照给定的精度把数字四舍五入。这有助于统计由于四舍五入产生的错误和其他由于浮点运算产生的问题
assertNotAlmostEqual(a, b)	round(a-b, 7) != 0	
assertGreater(a, b)	a > b	这些方法类似于 assertEqual()方法，是为逻辑判定条件设计的
assertGreaterEqual(a, b)	a >= b	
assertLess(a, b)	a < b	
assertLessEqual(a, b)	a <= b	
assertRegexpMatches(s, r)	r.search(s)	这些方法检查文本是否符合正则匹配
assertNotRegexpMatches(s, r)	not r.search(s)	
assertMultiLineEqual(a, b)	strings	此方法是 assertEqual()的一种特殊形式，为多行字符串设计。等值校验和其他单行字符串校验一样，但是默认失败信息经过优化以后可以展示具体值之间的差别
assertListEqual(a, b)	lists	此方法校验两个 list 是否相等，对于下拉列表选项字段的校验是非常有用的
fail()		此方法是无条件的失败。在别的 assert 方法不好用的时候，也可用此方法来创建定制的条件块

2.1.4　测试套件

应用 unittest 的 TestSuites 特性，可以将不同的测试组成一个逻辑组，然后设置统一的测试套件，并通过一个命令来执行测试。这都是通过 TestSuites、TestLoader 和 TestRunner 类来实现的。

在了解 TestSuites 的细节之前，我们为例子程序添加一个新的测试，用于校验主页。我们将把新加的测试和之前的测试放到一个测试组件中，详见下面代码。

```python
import unittest
from selenium import webdriver
from selenium.common.exceptions import NoSuchElementException
from selenium.webdriver.common.by import By
from __builtin__ import classmethod

class HomePageTest(unittest.TestCase):
    @classmethod
    def setUp(cls):
        # create a new Firefox session """
        cls.driver = webdriver.Firefox()
        cls.driver.implicitly_wait(30)
        cls.driver.maximize_window()
        # navigate to the application home page """
        cls.driver.get("http://demo.magentocommerce.com/")

    def test_search_field(self):
        # check search field exists on Home page
        self.assertTrue(self.is_element_present(By.NAME,"q"))

    def test_language_option(self):
        # check language options dropdown on Home page
        self.assertTrue(self.is_element_present
          (By.ID,"select-language"))

    def test_shopping_cart_empty_message(self):
        # check content of My Shopping Cart block on Home page
        shopping_cart_icon = \
            self.driver.find_element_by_css_selector
               ("div.header-minicart span.icon")
        shopping_cart_icon.click()
```

```python
        shopping_cart_status = \
            self.driver.find_element_by_css_selector\
                ("p.empty").text
        self.assertEqual("You have no items in your shopping cart.", shopping_cart_status)

        close_button = self.driver.find_element_by_css_selector\
            ("div.minicart-wrapper a.close")
        close_button.click()

    @classmethod
    def tearDown(cls):
        # close the browser window
        cls.driver.quit()

    def is_element_present(self, how, what):
        """
        Utility method to check presence of an element on page
        :params how: By locator type
        :params what: locator value
        """
        try: self.driver.find_element(by=how, value=what)
        except NoSuchElementException, e: return False
        return True

if __name__ == '__main__':
    unittest.main(verbosity=2)
```

我们将用 TestSuite 类来定义和执行测试套件。我们可以把多个测试加入一个测试套件中去。除了 TestSuite 类，我们还可以用 TestLoader 和 TextTestRunner 来创建和运行测试套件，举例如下。

```python
import unittest
from searchtests import SearchTests
from homepagetests import HomePageTest


# get all tests from SearchProductTest and HomePageTest class
```

```
search_tests = unittest.TestLoader().loadTestsFromTestCase
(SearchTests)
home_page_tests = unittest.TestLoader().loadTestsFromTestCase
(HomePageTest)

# create a test suite combining search_test and home_page_test
smoke_tests = unittest.TestSuite([home_page_tests, search_tests])

# run the suite
unittest.TextTestRunner(verbosity=2).run(smoke_tests)
```

使用 TestLoader 类，我们将得到指定测试文件中的所有测试方法且用于创建测试套件。TestRunner 类将通过调用测试套件来执行文件中所有的测试。

我们可以通过下面的命令运行新的测试套件文件。

```
python smoketests.py
```

这将运行 SearchProductTest 类和 HomePageTest 类中的所有测试并且通过命令行形式生成下图这样的测试输出。

```
C:\setests\chapter2>smoketests.py
test_language_option (homepagetests.HomePageTest) ... ok
test_search_field (homepagetests.HomePageTest) ... ok
test_shopping_cart_empty_message (homepagetests.HomePageTest) ... ok
test_search_by_category (searchtests.SearchTests) ... ok
test_search_by_name (searchtests.SearchTests) ... ok
----------------------------------------------------------------------
Ran 5 tests in 122.138s

OK
```

2.2 生成 HTML 格式的测试报告

unittest 在命令行输出测试结果。你可能需要生成一个所有测试的执行结果作为报告或者把测试结果发给相关人员。给相关人员发送命令行日志不是一个明智的选择。他们需要格式更加友好的测试报告，既能够查看测试结果的概况，也能够深入查看报告细节。unittest 没有相应的内置模块可以生成格式友好的报告，我们可以应用 Wai Yip Tung 编写的 unittest 的扩展 HTMLTestRunner 来实现。从 pypi.Python 的网站可以获取更多关于 HTMLTestRunner 的信息并可以下载说明文档。

 HTMLTestRunner 扩展可以在本书的附件源代码中找到。

我们将在测试中使用 HTMLTestRunner 来生成漂亮的测试报告。通过修改在本章前面涉及的测试套件文件来添加 HTMLTestRunner 支持。我们需要创建一个包含实际测试报告的输出文件，需要配置 HTMLTestRunner 选项和运行测试，具体如下。

```
import unittest
import HTMLTestRunner
import os
from searchtests import SearchTests
from homepagetests import HomePageTest

# get the directory path to output report file
dir = os.getcwd()

# get all tests from SearchProductTest and HomePageTest class
search_tests = unittest.TestLoader().loadTestsFromTestCase(SearchTests)
home_page_tests = unittest.TestLoader().loadTestsFromTestCase(HomePageTest)

# create a test suite combining search_test and home_page_test
smoke_tests = unittest.TestSuite([home_page_tests, search_tests])

# open the report file
outfile = open(dir + "\SmokeTestReport.html", "w")

# configure HTMLTestRunner options
runner = HTMLTestRunner.HTMLTestRunner(
            stream=outfile,
            title='Test Report',
            description='Smoke Tests'
            )
```

```
# run the suite using HTMLTestRunner
runner.run(smoke_tests)
```

执行该测试套件，HTMLTestRunner 像 unittest 的默认测试执行器一样运行所有的测试。在用例执行的最后，它将生成测试报告文件，如下图所示。

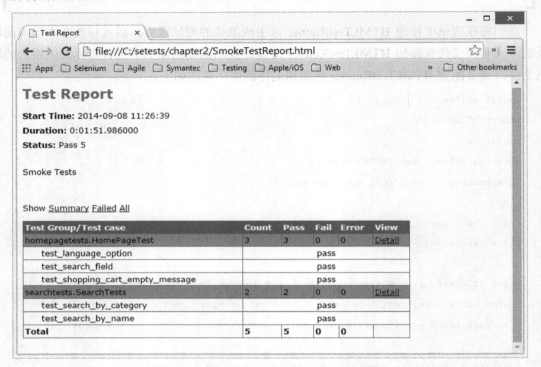

2.3 本章回顾

本章我们学习了如何使用 unittest 来编写和运行基于 Selenium WebDriver 的测试脚本。我们使用包含 setUp()方法和 tearDown()方法的 TestCase 类来创建测试。我们还可以添加断言来验证预期结果和实际结果是否一致。

我们也学习了如何使用 unittest 支持的不同类型的断言。我们实现了测试套件，它提供了把不同的测试用例组成逻辑分组的能力。最后，我们使用 HTMLTestRunner 来生成格式友好的 HTML 格式的测试报告。

在下一章中，我们将学习如何定义和使用定位器来与页面中不同类型的 HTML 元素进行交互。

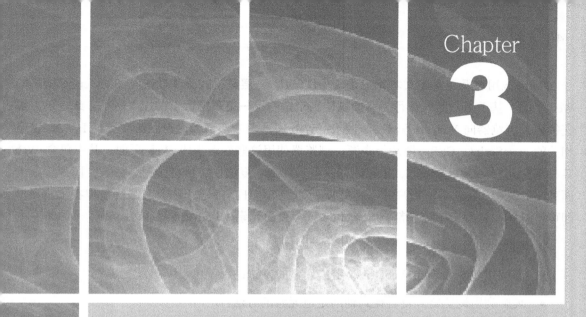

第 3 章
元素定位

Web 应用以及包含**超文本标记语言（HTML）**、**层叠样式表（CSS）**、JavaScript 脚本的 Web 页面；基于用户的操作行为诸如跳转到指定的**统一资源定位（URL）**网站，或是单击提交按钮，浏览器向 Web 服务器发送请求；Web 服务器响应请求，返回给浏览器 HTML 以及相关的 JavaScript、CSS、图片等资源；浏览器使用这些资源生成 Web 页面，其中包含 Web 各种视觉元素，例如文本框、按钮、标签页、图表、复选框、单选按钮、列表、图片等。上面列举的这些，我们普通用户并不用关心，放心交给 HTML 去组织并且最终呈现在浏览器里即可。这些视觉元素或控件都被 Selenium 称为**页面元素（WebElements）**。

本章包含以下主题：

- 理解更多 Selenium WebDriver 定位元素的方法；
- 理解利用浏览器开发者模式辅助定位元素的方法；
- 定位元素的方法包括通过 ID 定位、name 定位、class 属性定位以及利用 XPath 和 CSS 选择器定位；
- 通过各种 find_element_by 的方法查找元素，以便使用 Selenium WebDriver 与之自动化交互。

当我们想让 Selenium 自动地操作我们的浏览器，就必须告诉 Selenium 如何去定位某个元素或一组元素，可以通过编程的方式去模拟用户操作。每个元素有着不同的标签名和属性值，Selenium 提供了多种选择与定位元素的方法。

我们如何获取这些信息呢？大家都知道，Web 页面是由 HTML、CSS 和 JavaScript 等组成的。我们可以通过查看页面源文件的方式了解这些文本信息，进而可以找到我们想要的 tag 标签，了解与之对应的元素是如何交互的、如何定义属性与属性值的，以及页面的结构。下面展示了一个我们正在测试的场景，这是一个常见的搜索功能，包括搜索框与搜索按钮（放大镜图标）。

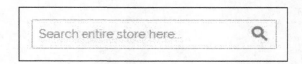

让我们看一下其对应的 HTML 脚本。

```
<form id="search_mini_form" action=
  "http://demo.magentocommerce.com/catalogsearch/result/"
  method="get">
```

```
        <div class="form-search">
            <label for="search">Search:</label>
            <input id="search" type="text" name="q" value=""
              class="input-text" maxlength="128" />
            <button type="submit" title="Search"
              class="button"><span><span>Search</span></span></button>
            <div id="search_autocomplete" class="search-
              autocomplete"></div>
            <script type="text/javascript">
            //<![CDATA[
                var searchForm = new Varien.searchForm
                  ('search_mini_form', 'search', 'Search entire store
                  here...');
                searchForm.initAutocomplete
                  ('http://demo.magentocommerce.com
                  /catalogsearch/ajax/suggest/',
                  'search_autocomplete');
            //]]>
            </script>
        </div>
</form>
```

我们发现类似搜索框、搜索按钮这样的元素，都是采用内嵌在<form>标签内的<input>标签来实现的，标记则用了<label>标签来实现。另外，JavaScript 代码写在<script>标签内。

其中搜索框<input>标签中包含 id、type、name、value、class 和 maxlength 属性的定义。

```
<input id="search" type="text" name="q" value=""
    class="input-text" maxlength="128" />
```

我们可以在浏览器窗口右键单击，在快捷菜单中选择**查看源文件**选项，在弹出的窗口中可以显示 HTML 文件与 JavaScript 脚本。

> 如果你对查看 HTML、CSS 和 JavaScript 感到生疏的话，可以查看相关网站，可以帮助你更快地识别 WebDriver 所需的元素的位置。

3.1 借助浏览器开发者模式定位

在使用 Selenium 测试之前，我们通常会先去查看页面源代码，借助工具可以帮助我们了解页面结构。值得庆幸的是，目前绝大多数的浏览器都内置有相关插件，能够快速、简洁地展示各类元素的属性定义、DOM 结构、JavaScript 代码块、CSS 样式等属性。接下来我们一起学习这类工具的细节以及使用方法。

3.1.1 用火狐浏览器 Firebug 插件检查页面元素

较新版本的火狐浏览器尽管自带有页面分析工具，然而，我们还是建议大家使用功能更强大的 Firebug 插件。

（1）你可以通过 addons.mozilla 网站，下载并安装 Firebug 插件。

（2）尝试用 Firebug，在页面上移动鼠标至希望获取的元素，然后右键单击，弹出快捷菜单。

（3）选择**使用 Firebug 查看元素**。此时火狐浏览器下方会显示 HTML 代码树窗口并定位到所选的元素上，如下图所示。

(4)我们还可以使用 Firebug 的 XPath 或 CSS 选择器,通过 Firebug 弹出窗口自带的检索功能,只要键入想要查找的关键字,就可以高亮显示与之匹配的元素,如下图所示。

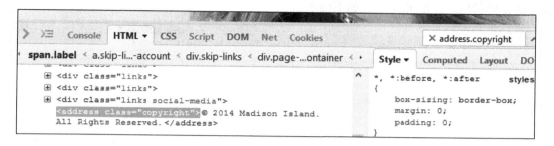

3.1.2 用谷歌 Chrome 浏览器检查页面元素

谷歌 Chrome 浏览器也自带有页面分析的功能。你可以通过以下步骤来检查页面元素。

(1)首先移动鼠标光标到期望的元素上,然后右键单击,在弹出的快捷菜单中,选择**检查(N)** 选项。

在浏览器的下方,将显示类似 Firebug 的**开发者工具**窗口,如下图所示。

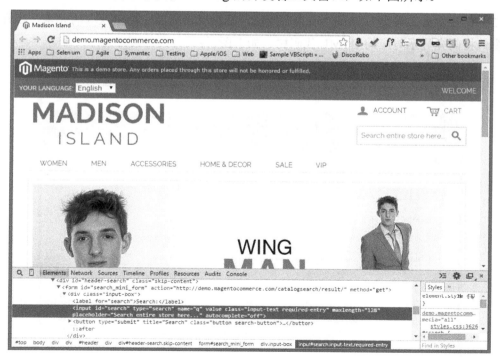

（2）同样类似于 Firebug，当我们需要使用 XPath 或 CSS 选择器时，在开发者工具下的 Elements 窗口中，按 Ctrl+F 键，将会显示搜索框。一旦输入 XPath 或 CSS 表达式，Firebug 就会高亮显示与之匹配的元素，如下图所示。

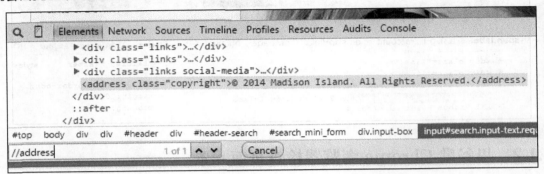

3.1.3 用 IE 浏览器检查页面元素

微软公司的 IE 浏览器也自带有页面分析的功能。你可以通过以下步骤来检查页面元素。

（1）按 F12 键，在浏览器下方显示**开发者工具窗口**。

（2）在**开发者工具窗口**选择"箭头"按钮，然后单击页面中期望获取的元素，在**开发者工具窗口**将高亮显示对应的 HTML 代码树，如下图所示。

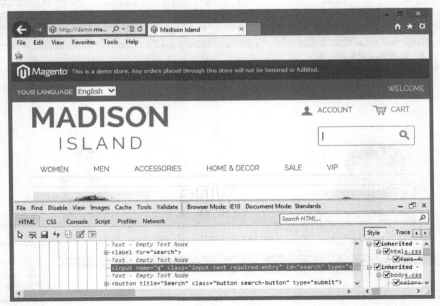

上述工具可以非常有效地帮助我们编写测试代码,以及执行与调试 JavaScript 脚本。

3.2 元素定位

我们必须告诉 Selenium 怎样去定位元素,用来模拟用户动作,或者查看元素的属性和状态,以便我们可以执行检查。例如,我们要搜索一个产品,首先要找到搜索框与搜索按钮,接着通过键盘输入要查询的关键字,最后用鼠标单击搜索按钮,提交搜索请求。

正如上述人工的操作步骤一样,我们也希望 Selenium 能模拟我们的动作,然而,Selenium 并不能理解类似在搜索框中输入关键字或单击搜索按钮这样图形化的操作。所以需要我们程序化地告诉 Selenium 如何定位搜索框与搜索按钮,从而模拟键盘与鼠标的动作。

Selenium 提供多种 find_element_by 方法用于定位页面元素。这些方法根据一定的标准去查找元素,如果元素被正常定位,那么 WebElement 实例将返回。反之,将抛出 NoSuchElementException 的异常。同时,Selenium 还提供多种 find_elements_by 方法去定位多个元素,这类方法根据所匹配的值,搜索并返回一个 list 数组(元素)。

Selenium 提供了 8 种 find_element_by 方法用于定位元素。接下来的部分,我们将逐一介绍方法细节,如下表所示。

方法	描述	参数	示例
find_element_by_id(id)	通过元素的 ID 属性值来定位元素	id:元素的 ID	driver.find_element_by_id('search')
find_element_by_name(name)	通过元素的 name 属性值来定位元素	name:元素的 name	driver.find_element_by_name('q')
find_element_by_class_name(name)	通过元素的 class 名来定位元素	name:元素的类名	driver.find_element_by_class_name('input-text')
find_element_by_tag_name(name)	通过元素的 tag name 来定位元素	name:tag name	driver.find_element_by_tag_name('input')
find_element_by_xpath(xpath)	通过 XPath 来定位元素	XPath:元素的 XPath	driver.find_element_by_xpath('//form[0]/div[0]/input[0]')
find_element_by_css_selector(css_selector)	通过 CSS 选择器来定位元素	css_selector:元素的 CSS 选择器	driver.find_element_by_css_selector('#search')
find_element_by_link_text(link_text)	通过元素标签对之间的文本信息来定位元素	link_text:文本信息	driver.find_element_by_link_text('Log In')
find_element_by_partial_link_text(link_text)	通过元素标签对之间的部分文本信息来定位元素	link_text:部分文本信息	driver.find_element_by_partial_link_text('Log')

find_elements_by 方法能按照一定的标准返回一组元素，具体见下表。

方法	描述	参数	示例
find_elements_by_id(id_)	通过元素的 ID 属性值来定位一组元素	id_：元素的 ID	driver.find_elements_by_id('product')
find_elements_by_name(name)	通过元素的 name 属性值来定位一组元素	name：元素的 name	driver.find_elements_by_name('products')
find_elements_by_class_name(name)	通过元素的 class 名来定位一组元素	name：元素的类名	driver.find_elements_by_class_name('foo')
find_elements_by_tag_name(name)	通过元素的 tag name 来定位一组元素	name：tag name	driver.find_elements_by_tag_name('a')
find_elements_by_xpath(xpath)	通过 XPath 来定位一组元素	XPath：元素的 XPath	driver.find_elements_by_xpath("//div[contains(@class,'lists')]")
find_elements_by_css_selector(css_selector)	通过 CSS 选择器来定位一组元素	css_selector：元素的 CSS 选择器	driver.find_elements_by_css_selector('.input-class')
find_elements_by_link_text(text)	通过元素标签对之间的文本信息来定位一组元素	text：文本信息	driver.find_elements_by_link_text('Log In')
find_elements_by_partial_link_text(link_text)	通过元素标签对之间的部分文本信息来定位一组元素	link_text：部分文本信息	driver.find_elements_by_partial_link_text('Add to,')

3.2.1　ID 定位

通过 ID 查找元素是查找页面上元素的最佳方法。find_element_by_id()和 find_elements_by_id()方法返回与 ID 属性值匹配的一个元素或一组元素。

find_element_by_id()方法返回的是与 ID 属性值匹配的第一元素，如果没有元素与之匹配，则抛出 NoSuchElementException 异常。

如下图所示，我们尝试来定位搜索框。

通过查看 HTML，我们可以看到搜索框的 ID 值被定义为 search。

```
<input id="search" type="text" name="q" value=""
  class="input-text" maxlength="128" autocomplete="off">
```

接下来我们使用 find_element_by_id() 方法，id 值为 search 来定位搜索框，同时检查 maxlength 的属性值。

```python
def test_search_text_field_max_length(self):
    # get the search textbox
    search_field = self.driver.find_element_by_id("search")

    # check maxlength attribute is set to 128
    self.assertEqual("128", search_field.get_attribute("maxlength"))
```

此外，如果使用 find_elements_by_id() 方法，那么将返回匹配 ID 值的所有元素。

3.2.2 name 定位

通过 name 定位是另外一个常用的查找元素的方式。find_element_by_name() 和 find_elements_by_name()方法可以通过匹配 name 值来定位单个或一组元素。同样，name 值匹配成功可返回定位的元素；反之，则抛出 NoSuchElementException 的异常。

回到之前的例子，我们可以用匹配 name 属性值的方式来替换 ID 值的匹配，同样可以定位到搜索框。

```python
# get the search textbox
self.search_field = self.driver.find_element_by_name("q")
```

此外，如果使用 find_elements_by_name() 方法，那么将返回匹配 name 值的所有元素。

3.2.3 class 定位

除了使用 ID 和 name 属性，我们还可通过 class 属性来定位元素。class 用来关联 CSS 中定义的属性。find_element_by_class_name()和 find_elements_by_class_name()方法可以通过匹配 class 属性来定位单个或一组元素。同样，class 值匹配成功可返回定位的元素；反之，则抛出 NoSuchElementException 的异常。

 通过对元素 ID、name 和 class 属性来查找元素是最为普遍和快捷的方法。此外，Selenium WebDriver 还提供了其他一些方法用于定位元素，在接下来的段落中会有详细介绍。

同样的场景（见下图），我们可以尝试用 find_element_by_class_name() 的方法来定位元素。

搜索按钮（放大镜图标）在 HTML 中是用<button>标签（元素）以及对应的 class 属性与属性值定义的，具体如下。

```
<button type="submit" title="Search"
  class="button"><span><span>Search</span></span></button>
```

下面我们用 class 属性值来定位搜索按钮，代码如下。

```
def test_search_button_enabled(self):
    # get Search button
    search_button = self.driver.find_element_by_class_name
        ("button")

    # check Search button is enabled
    self.assertTrue(search_button.is_enabled())
```

此外，如果使用 find_elements_by_class_name() 方法，那么将返回匹配 class 属性值的所有元素。

3.2.4 tag 定位

find_element_by_tag_name()和 find_elements_by_tag_name() 方法是通过对 HTML 页面中 tag name 匹配的方式来定位元素的。这些方法跟 JavaScript 中的 DOM 方法 getElementsByTagName()类似。同样，通过成功匹配 tag name 可以返回定位的元素；反之，则抛出 NoSuchElementException 的异常。

这个方法在某些特定的场景下格外有用，例如，我们可以通过<tr> 的 tag name 一次定位页面的 table 中所有的行数据（元素）。

在如下图所示的几张 banner 图中，通过查看 HTML 代码，我们可以看出包含有<img>或者<ul>的标签。

3.2 元素定位

这几张 banner 图采用无序列表标签<ul>内嵌图像标签<img>来实现。

```html
<ul class="promos">
    <li>
        <a href="http://demo.magentocommerce.com/home-decor.html">
            <img src="/media/wysiwyg/homepage-three-column-promo-
                01B.png" alt="Physical &amp; Virtual Gift Cards">
        </a>
    </li>
    <li>
        <a href="http://demo.magentocommerce.com/vip.html">
            <img src="/media/wysiwyg/homepage-three-column-promo-
                02.png" alt="Shop Private Sales - Members Only">
        </a>
    </li>
    <li>
        <a href="http://demo.magentocommerce.com/accessories/ bags-luggage.html">
            <img src="/media/wysiwyg/homepage-three-column-
                promo-03.png" alt="Travel Gear for Every Occasion">
        </a>
    </li>
</ul>
```

我们使用 find_elements_by_tag_name()方法来定位所有的 banner 图片。首先我们用 find_element_by_class_name()方法定位这一组 banner 图，然后用 find_elements_by_tag_name() 方法去匹配<img>的 tag name，最后把结果返回给 banners 对象。

```python
def test_count_of_promo_banners_images(self):
    # get promo banner list
    banner_list = self.driver.find_element_by_class_name("promos")

    # get images from the banner_list
    banners = banner_list.find_elements_by_tag_name("img")

    # check there are 20 tags displayed on the page
    self.assertEqual(2, len(banners))
```

3.2.5 XPath 定位

XPath 是一种在 XML 文档中搜索和定位元素的查询语言。几乎所有的浏览器都支持 XPath。同样，Selenium 也可以通过 XPath 的方式在 Web 页面上定位元素。

当我们发现通过 ID、name 或 class 属性值都无法定位元素时，不妨尝试用 XPath 的方式。我们可以灵活地运用绝对或相对路径定位，也可以通过除 ID、name 以外的其他属性来定位，甚至还可以通过属性值的一部分（如 starts-with()、contains()和 ends-with()）来帮助我们定位。

了解更多关于 XPath 的知识，可访问相关网站了解。

想要了解更多关于 XPath 定位的知识，可以参考《Selenium Testing Tools Cookbook》。

接下来，我们可以使用 find_element_by_xpath()和 find_elements_by_xpath() 方法来定位元素了。例如，我们通过之前的广告 banner 图，单击 banner 图片进入对应的页面。

上图是名为"SHOP PRIVATE SALES"的 banner 图，在<img> 的 tag 下，其中代码并不包含 ID、name 或 class 属性等信息，且这个页面还包含很多其他的<img>，所以我们不能通过传统的方法如 find_by_tag_name()简单地定位了。

```
<ul class="promos">
   ...
   <li>
      <a href="http://demo.magentocommerce.com/vip.html">
         <img src="/media/wysiwyg/homepage-three-column-
            promo-02.png" alt="Shop Private Sales - Members Only">
      </a>
   </li>
```

```
    ...
</ul>
```

我们尝试使用 find_element_by_xpath()方法，用<img>标签下的 alt 属性值来定位我们要找的元素。

代码如下。

```python
def test_vip_promo(self):
    # get vip promo image
    vip_promo = self.driver.\
        find_element_by_xpath("//img[@alt='Shop Private Sales - Members Only']")
    
    # check vip promo logo is displayed on home page
    self.assertTrue(vip_promo.is_displayed())
    # click on vip promo images to open the page
    vip_promo.click()
    # check page title
    self.assertEqual("VIP", self.driver.title)
```

此外，如果使用 find_elements_by_xpath() 方法，那么将返回匹配 XPath 查询到的所有元素。

3.2.6 CSS 选择器定位

CSS（层叠样式表）是一种用于页面设计（HTML）与表现的文件样式，是一种计算机语言，能灵活地为页面提供各种样式风格。CSS 使用选择器为页面元素绑定属性（如 ID、class、type、attribute、value 等）。

类似 XPath，Selenium 也可以利用 CSS 选择器的特性，帮助我们定位元素。如果想进一步了解关于 CSS 选择器的一些知识，请访问相关网站。

下面介绍 find_element_by_css_selector()和 find_elements_by_css_selector()两种方法。

回到首页的例子，可以看到购物车按钮，单击这个按钮，将进入购物车页面。如果此时没有添加任何商品，那么系统会提示"你还没有添加商品到购物车"，如下图所示。

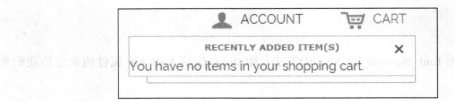

HTML 代码如下。

```
<div class="minicart-wrapper">
<p class="block-subtitle">
    Recently added item(s)
    <a class="close skip-link-close" href="#" title="Close">
      ×</a>
</p>
    <p class="empty">You have no items in your shopping cart.
      </p>
</div>
```

我们设计测试程序来校验这个提示信息。首先我们将使用 CSS 选择器来定位购物车按钮，然后单击它，紧接着定位即将弹出的信息。

```
def test_shopping_cart_status(self):
    # check content of My Shopping Cart block on Home page
    # get the Shopping cart icon and click to open the
    # Shopping Cart section
    shopping_cart_icon = self.driver.\
        find_element_by_css_selector("div.header-minicart span.icon")
    shopping_cart_icon.click()

    # get the shopping cart status
    shopping_cart_status = self.driver.\
        find_element_by_css_selector("p.empty").text
    self.assertEqual("You have no items in your shopping cart.", shopping_cart_status)
    # close the shopping cart section
    close_button = self.driver.\
        find_element_by_css_selector("div.minicart-wrapper a.close")
    close_button.click()
```

从上面的测试脚本可以看出，我们使用了元素 tag 和 class name 来缩小获取购物车按钮的范围。

```
shopping_cart_icon = self.driver.\
        find_element_by_css_selector("div.header-minicart span.icon")
```

首先定位到 tag 名为<div>的元素，然后接着".header-minicart"的类名，其下面的标签<span>下又有".icon"的类名。

想要了解更多关于 CSS 选择器的知识，请参考《Selenium Testing Tools Cookbook》。

3.2.7　Link 定位

find_element_by_link_text() 和 find_elements_by_link_text()方法是通过文本链接来定位元素。示例如下。

（1）定位首页上的 Account 链接，如下图所示，我们可以使用 find_element_by_link_text()方法。

（2）查看对应的 HTML 代码，具体如下。

```
<a href="#header-account" class="skip-link skip-account">
  <span class="icon"></span>
  <span class="label">Account</span>
</a>
```

（3）编写测试脚本，先通过文本定位 Account 链接，然后单击查看是否能显示。

```
def test_my_account_link_is_displayed(self):
    # get the Account link
    account_link = \
        self.driver.find_element_by_link_text("ACCOUNT")

    # check My Account link is displayed/visible in
    # the Home page footer
```

```
        self.assertTrue(account_link.is_displayed())
```

此外，如果使用 find_elements_by_link_text()方法，那么将返回匹配文本的所有元素。

3.2.8 Partial link 定位

find_element_by_partial_link_text()和 find_elements_by_partial_link_text()两个方法是通过文本链接的一部分文本来定位元素的方法。如下示例。

（1）同样是首页，有两个链接可以查看个人账户（Account）页面，一个是页面标头（header）部分的 Account 文字链接，另外一个是页脚（footer）部分的 My Account 文字链接。

（2）我们使用 find_elements_by_partial_link_text()方法，通过部分文本信息"Account"来定位，验证页面中的两个文本链接是否都能定位到（断言）。代码如下。

```
def test_account_links(self):
    # get the all the links with Account text in it
    account_links = self.driver.\
        find_elements_by_partial_link_text("ACCOUNT")

    # check Account and My Account link is displayed/visible in the Home page footer
    self.assertTrue(2, len(account_links))
```

3.3 方法实践

通过前面的介绍，我们尝试了很多种关于 find_element_by 的方法。接下来我们把这种类型的定位方法集成到同一个测试脚本中来。

（1）创建一个名为 homepagetest.py 的 Python 脚本，整合之前我们创建的那些测试代码。

```
import unittest
from selenium import webdriver

class HomePageTest(unittest.TestCase):
    @classmethod
    def setUpClass(cls):
```

```python
        # create a new Firefox session
        cls.driver = webdriver.Firefox()
        cls.driver.implicitly_wait(30)
        cls.driver.maximize_window()

        #navigate to the application home page
        cls.driver.get('http://demo.magentocommerce.com/')

    def test_search_text_field_max_length(self):
        # get the search textbox
        search_field = self.driver.
            find_element_by_id("search")

        # check maxlength attribute is set to 128
        self.assertEqual("128", search_field.get_attribute
        ("maxlength"))

    def test_search_button_enabled(self):
        # get Search button
        search_button = self.driver.
            find_element_by_class_name("button")

        # check Search button is enabled
        self.assertTrue(search_button.is_enabled())

    def test_my_account_link_is_displayed(self):
        # get the Account link
        account_link =
            self.driver.find_element_by_link_text("ACCOUNT")

        # check My Account link is displayed/visible in
        # the Home page footer
        self.assertTrue(account_link.is_displayed())

    def test_account_links(self):
```

```python
        # get the all the links with Account text in it
        account_links = self.driver.\
           find_elements_by_partial_link_text("ACCOUNT")

        # check Account and My Account link is
        # displayed/visible in the Home page footer
        self.assertTrue(2, len(account_links))

    def test_count_of_promo_banners_images(self):
        # get promo banner list
        banner_list = self.driver.\
           find_element_by_class_name("promos")

        # get images from the banner_list
        banners = banner_list.\
           find_elements_by_tag_name("img")

        # check there are 3 banners displayed on the page
        self.assertEqual(2, len(banners))

    def test_vip_promo(self):
        # get vip promo image
        vip_promo = self.driver.\
            find_element_by_xpath("//img[@alt=
            'Shop Private Sales - Members Only']")

        # check vip promo logo is displayed on home page
        self.assertTrue(vip_promo.is_displayed())
        # click on vip promo images to open the page
        vip_promo.click()
        # check page title
        self.assertEqual("VIP", self.driver.title)

    def test_shopping_cart_status(self):
        # check content of My Shopping Cart block
```

```python
        # on Home page
        # get the Shopping cart icon and click to
        # open the Shopping Cart section
        shopping_cart_icon = self.driver.\
            find_element_by_css_selector("div.header-
                minicart span.icon")
        shopping_cart_icon.click()

        # get the shopping cart status
        shopping_cart_status = self.driver.\
            find_element_by_css_selector("p.empty").text
        self.assertEqual("You have no items in your shopping
        cart.", shopping_cart_status)
        # close the shopping cart section
        close_button = self.driver.\
            find_element_by_css_selector("div.minicart-
                wrapper a.close")
        close_button.click()

    @classmethod
    def tearDownClass(cls):
        # close the browser window
        cls.driver.quit()

if __name__ == '__main__':
    unittest.main(verbosity=2)
```

（2）保存 py 文件，可以在命令行中直接执行。

```
python homepagetest.py
```

（3）执行测试过程中，unittest 会打印 7 组测试的执行结果（OK），如下图所示。

```
c:\setests\chapter3>homepagetests.py
test_account_links (__main__.HomePageTest) ... ok
test_count_of_promo_banners_images (__main__.HomePageTest) ... ok
test_my_account_link_is_displayed (__main__.HomePageTest) ... ok
test_search_button_enabled (__main__.HomePageTest) ... ok
test_search_text_field_max_length (__main__.HomePageTest) ... ok
test_shopping_cart_status (__main__.HomePageTest) ... ok
test_vip_promo (__main__.HomePageTest) ... ok

----------------------------------------------------------------------
Ran 7 tests in 36.347s

OK
```

3.4 章节回顾

在本章，我们一起学习了许多重要的 Selenium 定位元素的方法。使用 find_element_by 方法，通过 ID、name、class name、tag name、XPath、CSS 选择器以及文本链接（或部分）去定位元素。

在后续设计测试时，可以更灵活地去运用这些定位的策略。这些知识为接下来的章节介绍的如何调用 Selenium API 奠定了基础。

下一章，将会学习如何使用 Selenium WebDriver 的功能去与定位到的元素交互，以及模拟用户的操作（例如，在文本框输入、单击按钮、选择下拉菜单、调用 JavaScript 等操作）。

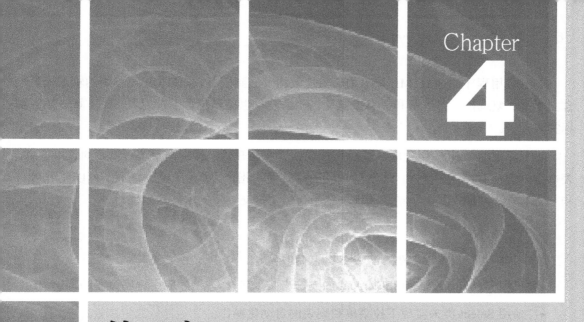

第 4 章
Selenium Python API 介绍

Web 应用程序通过 HTML 表单的形式把数据发送到服务端。HTML 表单中的输入元素包含文本框、复选框、单选框和提交按钮等。一个表单也可以包含下拉列表、文本域、插图和标签等元素。

一个典型的 Web 应用程序从注册用户或搜索产品开始，往往需要填写很多的表单信息。表单是内嵌在 HTML 代码的<form>标签里的。标签中指定了提交数据的方法，可以使用 GET 和 POST 方法，输入到表单请求的地址就是我们要提交数据的服务器地址。

本章包含以下主题：

- 更多地了解 WebDriver 和 WebElement 这两个类；
- 使用 WebDriver 和 WebElement 的方法来实现包含与 Web 应用程序交互的测试；
- 使用 Select 类来实现下拉菜单和列表的自动化操作；
- 实现 JavaScript 警告和浏览器导航栏的自动化。

4.1　HTML 表单元素

HTML 表单是由不同类型的元素组成的，如下图所示包含<fORM>、<INPUT>等元素。Web 应用开发者通过使用这些元素来实现数据的展示和接收用户的数据提交。开发人员通过定义这些元素来编写 Web 页面的 HTML 代码。然而作为终端用户，我们看到的这些元素是通过诸如文本框、标签、按钮、复选框和单选按钮的形式展现出来的。HTML 代码对于终端用户来说是不可见的。

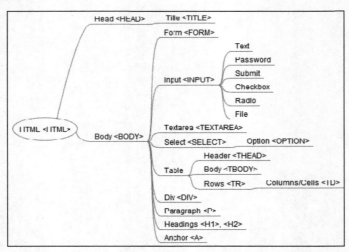

Selenium WebDriver 为"实现通过与这些 Web 元素的交互的自动化来验证 Web 程序功能的正确性"提供了强大的支持。

4.2 WebDriver 原理

WebDriver 提供许多用来与浏览器交互的功能和设置。我们可以通过使用 WebDriver 的功能和一些方法来实现与浏览器窗口、警告、框架和弹出窗口的交互。它也提供了自动化操作浏览器导航栏、设置 cookies、截屏等方便我们测试的特性。在后面的章节中，我们将依次阐述这些 WebDriver 的重要特性。本节的表格中包含了一些将在本书后面章节中使用到的非常重要的功能和方法。

在下面的网址可以看到完整的 WebDriver 的 properties 和方法列表。
http://selenium.googlecode.com/git/docs/api/py/webdriver_remote/selenium.webdriver.remote.webdriver.html#moduleselenium.webdriver.remote.webdriver

4.2.1 WebDriver 功能

WebDriver 通过下表的功能来操纵浏览器。

功能/属性	描述	实例
current_url	获取当前页面的 URL 地址	driver.current_url
current_window_handle	获取当前窗口的句柄	driver.current_window_handle
name	获取该实例底层的浏览器名称	driver.name
orientation	获取当前设备的方位	driver.orientation
page_source	获取当前页面的源代码	driver.page_source
title	获取当前页面的标题	driver.title
window_handles	获取当前 session 里所有窗口的句柄	driver.window_handles

4.2.2 WebDriver 方法

WebDriver 通过一些方法来实现与浏览器窗口、网页和页面元素的交互。下表是一些重要的方法。

方法	描述	参数	实例
back()	后退一步到当前会话的浏览器历史记录中最后一步操作前的页面		driver.back()
close()	关闭当前浏览器窗口		driver.close()
forward()	前进一步到当前会话的浏览器历史记录中前一步操作后的页面		driver.forward()
get(url)	访问目标 URL 并加载网页到当前的浏览器会话	URL 是目标网页的网站地址	driver.get("http://www.google.com")
maximize_window()	最大化当前浏览器窗口		driver.maximize_window()
quit()	退出当前 driver 并且关闭所有的相关窗口		driver.quit()
refresh()	刷新当前页面		driver.refresh()
switch_to_active_element()	返回当前页面唯一焦点所在的元素或者元素体		driver.switch_to_active_element()
switch_to_alert()	把焦点切换至当前页面弹出的警告		driver.switch_to_alert()
switch_to_default_content()	切换焦点至默认框架内		driver.switch_to_default_content()
switch_to-frame(frame_reference)	通过索引、名称和网页元素将焦点切换到指定的框架，这种方法也适用于 IFRAMES	frame_reference：要切换的目标窗口的名称、整数类型的索引或者要切换的目标框架的网页元素	driver.switch_to_frame ('frame_name')
switch_to_window (window_name)	切换焦点到指定的窗口	window_name：要切换的目标窗口的名称或者句柄	driver.switch_to_window ('main')
implicitly_wait(time_to_wait)	超时设置，等待目标元素被找到，或者目标指令执行完成。该方法在每个 session 只需要调用一次。execute_async_script 的超时设置，请参阅 set_script_timeout 方法	time_to_wait：等待时间（单位为秒）	
set_page_load_timeout (time_to_wait)	设置一个页面完全加载完成的超时等待时间	time_to_wait：等待时间（单位为秒）	driver.set_page_load_timeout(30)
set_script_timeout(time_to_wait)	设置脚本执行的超时时间，应该在 execute_async_script 抛出错误之前	time_to_wait：等待时间（单位为秒）	driver.set_script_timeout(30)

4.3 WebElement 接口

我们可以通过 WebElement 实现与网站页面上的元素的交互。这些元素包含文本框、文本域、按钮、单选框、多选框、表格、行、列和 div 等。

WebElement 提供了一些功能、属性和方法来实现与网页元素的交互。本节的表格中将列出后面章节会用到的一些重要的功能和方法。如果想查看完整的功能和方法详情，请访问以下网址。

selenium.googlecode/git/docs/api/py/webdriver_remote/selenium.webdriver.remote.webelement.html#module-selenium.webdriver.remote.webelement

4.3.1 WebElement 功能

下面是 WebElement 功能列表。

功能/属性	描述	实例
size	获取元素的大小	element.size
tag_name	获取元素的 HTML 标签名称	element.tag_name
text	获取元素的文本值	element.text

4.3.2 WebElement 方法

下面是 WebElement 方法列表。

方法	描述	参数	实例
clear()	清除文本框或者文本域中的内容		element.clear()
click()	单击元素		element.click()
get_attribute(name)	获取元素的属性值	name：元素的名称	element.get_attribute("value") 或者 element.get_attribute("maxlength")
is_displayed()	检查元素对于用户是否可见		element.is_displayed()

续表

方 法	描 述	参 数	实 例
is_enabled()	检查元素是否可用		element.is_enabled()
is_selected()	检查元素是否被选中。该方法应用于复选框和单选按钮		element.is_selected()
send_keys(*value)	模拟输入文本	value：待输入的字符串	element.send_keys("foo")
submit()	用于提交表单。如果对一个元素应用此方法，将会提交该元素所属的表单		element.submit()
value_of_css_property (property_name)	获取 CSS 属性的值	property_name：CSS 属性的名称	element.value_of_css_property ("backgroundcolor")

4.4 操作表单、文本框、复选框、单选按钮

我们可以使用 WebElement 实现与各种 HTML 控件的自动化交互，例如在一个文本框输入文本、单击一个按钮、选择单选按钮或者复选框、获取元素的文本和属性值等。

在前面的章节可以看到 WebElement 提供的功能和方法。在本节中，我们将使用 WebElement 及其功能和方法实现在样例程序中**创建账户**功能的自动化。接下来我们创建一个测试脚本，来验证被测程序是否能正确创建一个新的账户。我们将按照下图来填写表单信息并且提交请求，系统收到请求后应该创建一个新的账户。

正如在上图看到的,我们需要填写 5 个文本框并且选择一个复选框。

(1) 首先,创建一个新的测试类 RegisterNewUser,下面是实例代码。

```
from selenium import webdriver
import unittest


class RegisterNewUser(unittest.TestCase):
    def setUp(self):
        self.driver = webdriver.Firefox
        self.driver.implicitly_wait(30)
        self.driver.maximize_window()

        # navigate to the application home page
        self.driver.get("http://demo.magentocommerce.com/")
```

(2) 添加一个测试方法 test_register_new_user(self) 到 RegisterNewUser 类中。

(3) 为了打开登录页面,我们需要单击主页的登录链接。用于登录的代码如下。

```
def test_register_new_user(self):
    driver = self.driver

    # click on Log In link to open Login page
    driver.find_element_by_link_text("Log In").click()
```

4.4.1 检查元素是否启用或显示

当元素在屏幕上可见的时候(visible 属性设置为 TRUE),调用 is_displayed() 方法返回为 TRUE,反之就会返回 FALSE。类似地,当元素是可用的时候,调用 is_enabled() 方法返回为 TRUE,这时用户就可以执行点击和输入文本等操作。当元素是不可用的时候,该方法返回 FALSE。

用户登录页面提供了使用已有账户登录和创建新用户的选项。我们可以通过调用 is_displayed()方法和 is_enabled()方法检查创建新账户按钮对于用户是否可见并且可用。添加下面的代码到测试类中。

```
# get the Create Account button
```

```python
        create_account_button = driver.find_element_by_xpath("//button[@title='Create an Account']")

        # check Create Account button is displayed and enabled
        self.assertTrue(create_account_button.is_displayed() and
                        create_account_button.is_enabled())
```

我们要测试创建账户功能,因此要单击创建账户按钮,然后将会展示创建新账户的页面。我们可以通过检查 WebDriver 的 title 属性来校验打开的页面是否符合预期结果,代码如下。

```python
# click on Create Account button. This will display
# new account
create_account_button.click()


# check title
self.assertEquals("Create New Customer Account - Magento Commerce Demo Store", driver.title)
```

在**创建新账户**页面,可以通过调用 find_element_by_* 方法来查找定位所有的元素。

```python
# get all the fields from Create an Account form
first_name = driver.find_element_by_id("firstname")
last_name = driver.find_element_by_id("lastname")
email_address = driver.find_element_by_id("email_address")
news_letter_subscription =
  driver.find_element_by_id("is_subscribed")
password = driver.find_element_by_id("password")
confirm_password = driver.find_element_by_id("confirmation")
submit_button =
  driver.find_element_by_xpath("//button[@title='Submit']")
```

4.4.2 获取元素对应的值

get_attribute()方法可以用来获取元素的属性值。例如,单个测试是用来验证输入姓和名字的文本框的最大字符限制是 255,字符限制就是通过 maxlength 属性来实现的,如下代码所示设置值为 255。

```html
<input type="text" id="firstname" name="firstname" value=""
```

```
title="First Name" maxlength="255" class="input-text required-entry">
```

我们可以通过调用 get_attribute()方法来校验 maxlength 属性是否正确。

（1）需要把属性名称作为参数传递给 get_attribute()方法。

```
# check maxlength of first name and last name textbox
self.assertEqual("255", first_name.get_attribute("maxlength"))
self.assertEqual("255", last_name.get_attribute("maxlength"))
```

（2）添加以下代码到测试脚本中，以确保所有的字段对于用户都是可见和可用的。

```
# check all fields are enabled
        self.assertTrue(first_name.is_enabled()and last_name.is_enabled()
            and email_address.is_enabled() and news_letter_subscription.is_enabled()
            and password.is_enabled() and confirm_password.is_enabled()
            and submit_button.is_enabled())
```

4.4.3　is_selected()方法

is_selected() 方法是针对单选按钮和复选框的。我们可以通过调用该方法来得知一个单选按钮或复选框是否被选中。

单选按钮或复选框可以通过 WebElement 的 click() 方法来执行点击操作，从而选中该元素。如下面的例子，检查 Sign UP for Newsletter 复选框是否默认为不被选中的，示例代码如下。

```
# check Sign Up for Newsletter is unchecked
        self.assertFalse(news_letter_subscription.is_selected())
```

4.4.4　clear()与send_keys()方法

clear() 和 send_keys()方法适用于文本框和文本域，分别用于清除元素的文本内容和模拟用户操作键盘来输入文本信息。待输入的文本作为 send_keys() 方法的参数。

（1）添加下面的代码，通过 send_keys() 方法来给对应的字段填写值。

```
# fill out all the fields
first_name.send_keys("Test")
last_name.send_keys("User1")
news_letter_subscription.click()
```

```
email_address.send_keys("TestUser_150214_2200@example.com")
password.send_keys("tester")
confirm_password.send_keys("tester")
```

(2)最终通过校验欢迎信息来检查用户是否创建成功。

我们可以通过 text 属性来获取元素的文本内容。

```
# check new user is registered
self.assertEqual("Hello, Test User1!", driver.find_
    element_by_css_selector("p.hello > strong").text)
self.assertTrue(driver.find_element_by_link_text("LogOut").is_displayed())
```

(3)下面是创建一个账户功能的完整测试。运行这个测试脚本将看到在 Create An Account 页面的所有操作。

```
from selenium import webdriver
from time import gmtime, strftime
import unittest

class RegisterNewUser(unittest.TestCase):
    def setUp(self):
        self.driver = webdriver.Firefox()
        self.driver.implicitly_wait(30)
        self.driver.maximize_window()

        # navigate to the application home page
        self.driver.get("http://demo.magentocommerce.com/")

    def test_register_new_user(self):
        driver = self.driver

        # click on Log In link to open Login page
        driver.find_element_by_link_text("ACCOUNT").click()
        driver.find_element_by_link_text("My Account").click()

        # get the Create Account button
        create_account_button = \
```

```python
    driver.find_element_by_link_text("CREATE AN ACCOUNT")

# check Create Account button is displayed
# and enabled
self.assertTrue(create_account_button.
   is_displayed() and
   create_account_button.is_enabled())

# click on Create Account button. This will
# display new account
create_account_button.click()

# check title
self.assertEquals("Create New Customer Account", driver.title)

# get all the fields from Create an Account form
first_name = driver.find_element_by_id("firstname")
last_name = driver.find_element_by_id("lastname")
email_address = driver.find_element_by_id("email_address")
password = driver.find_element_by_id("password")
confirm_password = driver.find_element_by_id("confirmation")
news_letter_subscription = driver.find_element_by_id("is_subscribed")
submit_button = driver.\find_element_by_xpath ("//button[@title='Register']")

# check maxlength of first name and
# last name textbox
self.assertEqual("255", first_name.get_attribute("maxlength"))
self.assertEqual("255", last_name.get_attribute("maxlength"))

# check all fields are enabled
self.assertTrue(first_name.is_enabled()
   and last_name.is_enabled()
   and email_address.is_enabled() and
   news_letter_subscription.is_enabled() and
```

```python
            password.is_enabled() and
            confirm_password.is_enabled()
            and submit_button.is_enabled())

        # check Sign Up for Newsletter is unchecked
        self.assertFalse(news_letter_subscription. is_selected())

        user_name = "user_" + strftime ("%Y%m%d%H%M%S", gmtime())

        # fill out all the fields
        first_name.send_keys("Test")
        last_name.send_keys(user_name)
        news_letter_subscription.click()
        email_address.send_keys(user_name + "@example.com")
        password.send_keys("tester")
        confirm_password.send_keys("tester")

        # click Submit button to submit the form
        submit_button.click()

        # check new user is registered
        self.assertEqual("Hello, Test " + user_name + "!",
            driver.find_element_by_css_selector("p.hello >strong").text)
        driver.find_element_by_link_text("ACCOUNT").click()
        self.assertTrue(driver.find_element_by_link_text ("Log Out").is_displayed())

    def tearDown(self):
        self.driver.quit()

if __name__ == "__main__":
    unittest.main(verbosity=2)
```

4.5 操作下拉菜单

Selenium WebDriver 提供了特定的 Select 类实现与网页上的列表和下拉菜单的交互。例如下面的样例程序，可以看到一个为店铺选择语言的下拉菜单。

下拉菜单和列表是通过 HTML 的<select>元素实现的。选择项是通过<select>中的<option>元素实现的，如下 HTML 代码。

```
<select id="select-language" title="Your Language"
  onchange="window.location.href=this.value">
<option value="http://demo.magentocommerce.com/?
    ___store=default&amp;___from_store=default"
    selected="selected">English</option>
<option value="http://demo.magentocommerce.com/?
    ___store=french&amp;___from_store=default">French</option>
<option value="http://demo.magentocommerce.com/?
    ___store=german&amp;___from_store=default">German</option>
</select>
```

每个<option>元素都有属性值和文本内容，是用户可见的。例如，在下面的代码中，<option>设置的是店铺的 URL，后面参数设置的是语言种类，这里是 French。

```
<option value="http://demo.magentocommerce.com/customer/
  account/create/?___store=french&amp;
    ___from_store=default">French</option>
```

4.5.1 Select 原理

Select 类是 Selenium 的一个特定的类，用于与下拉菜单和列表交互。它提供了丰富的功能和方法来实现与用户交互。

下面两小节的表格列出来 Select 类中所有的功能和方法。你也可以在下面网址获取类似信息。

selenium.googlecode/git/docs/api/py/webdriver_support/selenium.webdriver.support.select.html#module- selenium.webdriver.support.select

4.5.2 Select 功能

Select 类实现的功能见下表。

功能/属性	描述	实例
all_selected_options	获取下拉菜单和列表中被选中的所有选项内容	select_element.all_selected_options
first_selected_option	获取下拉菜单和列表的第一个选项/当前选择项	select_element.first_selected_option
options	获取下拉菜单和列表的所有选项	select_element.options

4.5.3 Select 方法

Select 类实现的方法见下表。

功能/属性	描述	参数	实例
deselect_all()	清除多选下拉菜单和列表的所有选择项		select_element.deselect_all()
deselect_by_index(index)	根据索引清除下拉菜单和列表的选择项	index：要清除的目标选择项的索引	select_element.deselect_by_index(1)
deselect_by_value(value)	清除所有选项值和给定参数匹配的下拉菜单和列表的选择项	value：要清除的目标选择项的 value 属性	select_element.deselect_by_value("foo")
deselect_by_visible_text(text)	清除所有展示的文本和给定参数匹配的下拉菜单和列表的选择项	text：要清除的目标选择项的文本值	select_element.deselect_by_visible_text("bar")

续表

功能/属性	描述	参数	实例
select_by_index(index)	根据索引选择下拉菜单和列表的选择项	index：要选择的目标选择项的索引	select_element.select_by_index(1)
select_by_value(value)	选择所有选项值和给定参数匹配的下拉菜单和列表的选择项	value：要选择的目标选择项的value属性	select_element.select_by_value("foo")
select_by_visible_text(text)	选择所有展示的文本和给定参数匹配的下拉菜单和列表的选择项	text：要选择的目标选择项的文本值	select_element.select_by_visible_text("bar")

让我们进一步探究这些功能和方法，我们回到刚才被测网站的**语言选择**功能。我们将为前面章节创建好的主页面的测试类添加一个新的测试用例。这个测试用例用来验证是否有8种语言可供用户选择。我们将首先使用options属性来验证选项的个数是否和预期结果一致，然后通过获取每个选项的文本来与预期的选项列表相比较，从而校验是否一致，代码如下所示。

```python
def test_language_options(self):
    # list of expected values in Language dropdown
    exp_options = ["ENGLISH", "FRENCH", "GERMAN"]
    
    # empty list for capturing actual options displayed
    # in the dropdown
    act_options = []
    
    # get the Your language dropdown as instance of Select class
    select_language = \
        Select(self.driver.find_element_by_id("select-language"))
    
    # check number of options in dropdown
    self.assertEqual(2, len(select_language.options))
    
    # get options in a list
    for option in select_language.options:
        act_options.append(option.text)
    
    # check expected options list with actual options list
```

```
        self.assertListEqual(exp_options, act_options)

        # check default selected option is English
        self.assertEqual("ENGLISH", select_language.first_selected_option.text)

        # select an option using select_by_visible text
        select_language.select_by_visible_text("German")

        # check store is now German
        self.assertTrue("store=german" in self.driver.current_url)

        # changing language will refresh the page,
        # we need to get find language dropdown once again
        select_language = \
           Select(self.driver.find_element_by_id("select-language"))
           select_language.select_by_index(0)
```

options 属性返回一个下拉选项和列表里的所有 <option> 元素。选项列表里的每个选项都是一个 WebElement 类的实例。

我们也可以通过用 first_selected_ option 属性来校验默认/当前选择项是否正确。

 all_selected_options 属性是用来测试多选的下拉选项和列表的。

最后，我们用下面的代码来实现：选择一个语言选项，然后校验**保存的 URL** 是否能够随着语言选项的改变而正确地变化。

```
# select an option using select_by_visible text
select_language.select_by_visible_text("German")

# check store is now German
self.assertTrue("store=german" in self.driver.current_url)
```

一个或多个选项可以基于索引来选择（该选项在列表中的位置），也可以根据属性值或者文本值来选择。select 类提供了很多 select_ 方法来选择选项。在上面这个例子中，我们使用 select_by_visible_text() 方法来选择选项。反之，我们也可以用各种 deselect_ 方法来取消选择。

4.6 操作警告和弹出框

开发人员使用 JavaScript 警告或者模态对话框来提示校验错误信息、报警信息、执行操作后的返回信息,甚至用来接收输入值等。本节我们将了解如何使用 Selenium 来操控警告和弹出框。

4.6.1 Alert 原理

Selenium WebDriver 通过 Alert 类来操控 JavaScript 警告。Alert 包含的方法有接受、驳回、输入和获取警告窗口的文本。

4.6.2 Alert 功能

Alert 实现了下表的功能。

功能/属性	描述	实例
text	获取警告窗口的文本	alert.text

4.6.3 Alert 方法

Alert 实现了下表的方法。

方法	描述	参数	实例
accept()	接受 JavaScript 警告信息,单击 OK 按钮		alert.accept()
dismiss()	驳回 JavaScript 警告信息,单击取消按钮		alert.dismiss()
send_keys(*value)	模拟给元素输入信息	value:待输入目标字段的字符串	alert.send_keys("foo")

在样例程序中,可以看到使用了 Alert 通知或告警。用户先添加产品进行比较,然后移除一个或多个产品时,被测程序将会显示一个如下图这样的告警信息。

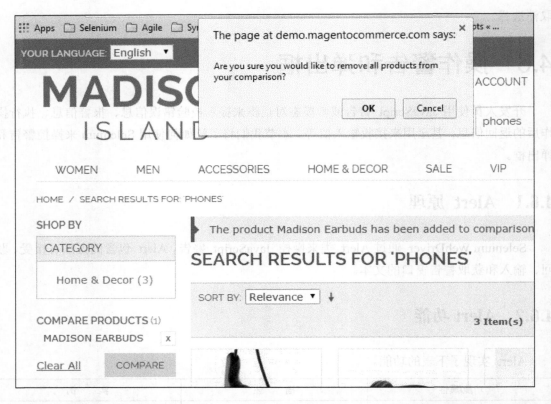

我们将设计一个测试来验证单击 COMPARE PRODUCTS（产品比较）功能中的 Clear All 链接时，是否会弹出警告提醒用户。

创建一个新的测试类 CompareProducts，并添加测试场景的代码，搜索并添加一个产品到比较列表中，代码如下。

```
from selenium import webdriver
import unittest


class CompareProducts(unittest.TestCase):
    def setUp(self):
        self.driver = webdriver.Firefox()
        self.driver.implicitly_wait(30)
        self.driver.maximize_window()
        self.driver.get("http://demo.magentocommerce.com/")
```

```python
def test_compare_products_removal_alert(self):
    # get the search textbox
    search_field = self.driver.find_element_by_name("q")
    search_field.clear()

    # enter search keyword and submit
    search_field.send_keys("phones")
    search_field.submit()

    # click the Add to compare link
    self.driver.\
        find_element_by_link_text("Add to Compare").click()
```

当单击 Add to Compare 链接将一个产品添加到比较列表时，将会看到一个产品被添加到 COMPARE PRODUCTS 下面。这个时候还可以添加其他的产品到比较列表。如果想从比较列表移除所有的产品，可以在 COMPARE PRODUCTS 模块中单击 Clear All 链接。这个时候可以看到一个警告提示"是否确认移除所有的产品"。我们可以通过 Alert 来操控这个警告。调用 WebDriver 的 Switch_to_alert() 方法可以返回一个 Alert 的实例。我们可以利用这个 Alert 实例来获取警告信息，并通过单击 OK 按钮来接受这个警告信息，或者通过单击 Cancel 按钮来拒绝这个警告。添加下面的代码到测试脚本中，这部分代码用来读取并且校验警告信息是否正确，然后通过调用 accept() 方法来接受警告。

```python
    # click on Remove this item link, this will display
    # an alert to the user
    self.driver.find_element_by_link_text("Clear All").click()

    # switch to the alert
    alert = self.driver.switch_to_alert()

    # get the text from alert
    alert_text = alert.text

    # check alert text
    self.assertEqual("Are you sure you would like to "
      "remove all products from your comparison?", alert_text)
```

```python
    # click on Ok button
    alert.accept()

    def tearDown(self):
        self.driver.quit()

if __name__ == "__main__":
    unittest.main()
```

4.6.4 浏览器自动化处理

通过单击浏览器工具栏上的后退、前进、刷新/重新加载按钮，可以实现访问历史页面、刷新当前页面等操作。Selenium WebDriver API 提供了很多操控这些按钮的方法，我们可以使用这些方法来验证浏览器的行为。WebDriver 类提供了以下方法来操控浏览器的后退、前进和刷新等操作。

方法	描述	参数	实例
back()	后退到浏览器当前会话的历史记录中的前一步操作	无	driver.back()
forward()	向前一步到浏览器当前会话的历史记录中的后一步操作	无	driver.forward()
refresh()	刷新浏览器中的当前页面	无	driver.refresh()

下面的例子是通过浏览器 API 操控浏览器的历史记录并验证程序的状态。

```python
import unittest
from selenium import webdriver
from selenium.webdriver.support.ui import WebDriverWait
from selenium.webdriver.support import expected_conditions

class NavigationTest(unittest.TestCase):
    def setUp(self):
        # create a new Firefox session
        self.driver = webdriver.Chrome()
        self.driver.implicitly_wait(30)
```

```python
    self.driver.maximize_window()

    # navigate to the application home page
    self.driver.get("http://www.google.com")

def testBrowserNavigation(self):
    driver = self.driver
    # get the search textbox
    search_field = driver.find_element_by_name("q")
    search_field.clear()

    # enter search keyword and submit
    search_field.send_keys("selenium webdriver")
    search_field.submit()

    se_wd_link = driver.find_element_by_link_text("Selenium WebDriver")
    se_wd_link.click()
    self.assertEqual("Selenium WebDriver", driver.title)

    driver.back()
    self.assertTrue(WebDriverWait(self.driver, 10)
        .until(expected_conditions.title_is
            ("selenium webdriver - Google Search")))

    driver.forward()
    self.assertTrue(WebDriverWait(self.driver, 10)
        .until(expected_conditions.title_is
            ("Selenium WebDriver")))

    driver.refresh()
    self.assertTrue(WebDriverWait(self.driver, 10)
      .until(expected_conditions.title_is
        ("Selenium WebDriver")))
```

```
    def tearDown(self):
        # close the browser window
        self.driver.quit()

if __name__ == '__main__':
    unittest.main()
```

4.7　章节回顾

本章介绍了 Selenium WebDriver API 与页面各种元素的交互实现。Selenium WebDriver API 提供了不同的类、功能和方法来模拟用户的动作，从而校验应用程序的状态。这些方法能够自动化操控的元素有文本框、按钮、复选框和下拉列表等。

同时，我们还设计了一些处理警告的测试，学习了操控浏览器的方法，并且测试了浏览器在不同页面之间的跳转。

在下一章，我们将进一步学习 Selenium API 如何来处理同步机制，这些内容能够帮助我们构建更加稳定的测试。

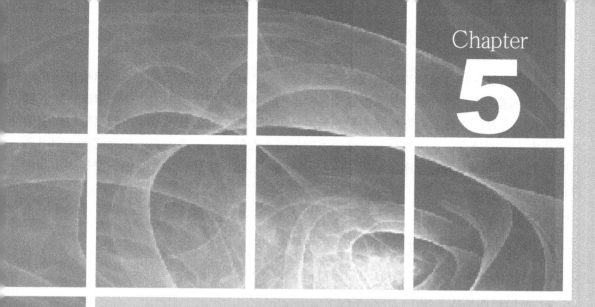

第 5 章
元素等待机制

能否构建健壮和可靠的测试是 UI 自动化测试能否成功的关键因素之一。然而当一个测试接着一个测试执行的时候，常常会遇到各种不同的状况。当使用脚本定位元素或去验证程序的运行状态时，有时候会发现找不到元素，这可能是由突然的资源受限或网络延迟引起的响应速度太慢所导致，这时测试报告就会返回测试失败的结果。我们需要在测试脚本中引入延时机制，来使脚本的运行速度与程序的响应速度相匹配。换句话说，我们需要使脚本和程序的响应能够同步。WebDriver 为这种同步提供了隐式等待和显式等待两种机制。

本章包含以下主题：

- 如何使用隐式等待或显式等待？
- 什么情况下使用隐式等待或显式等待？
- 使用预期等待条件；
- 创建自定义的等待条件。

5.1 隐式等待

隐式等待为 WebDriver 中的完整的一个测试用例或者一组测试的同步，提供了通用的方法。隐式等待对于解决由于网络延迟或利用 Ajax 动态加载元素所导致的程序响应时间不一致问题，是非常有效的。

当设置了隐式等待时间后，WebDriver 会在一定的时间内持续检测和搜寻 DOM，以便于查找一个或多个不是立即加载成功并可用的元素。一般情况下，隐式等待的默认超时时间设置为 0 秒。

一旦设置，隐式等待时间就会作用于这个 WebDriver 实例的整个生命周期或者一次完整测试的执行期间。

WebDriver 类提供了 implicitly_wait()方法来配置超时时间。本书在第 2 章已经创建了 SearchProductTest 测试类，是基于 unittest 写的测试。我们将基于这个类进行修改，在 setUp() 方法中加入隐式等待时间并且设置为 10 秒，代码如下面的例子所示。当一个测试用例执行的时候，WebDriver 在找不到一个元素的时候，将会等待 10 秒。当达到 10 秒超时时间后，将会抛出一个 NoSuchElementException 的异常。

```
import unittest
```

```python
from selenium import webdriver

class SearchProductTest(unittest.TestCase):
    def setUp(self):
        # create a new Firefox session
        self.driver = webdriver.Firefox()
        self.driver.implicitly_wait(30)
        self.driver.maximize_window()

        # navigate to the application home page
        self.driver.get("http://demo.magentocommerce.com/")

    def test_search_by_category(self):
        # get the search textbox
        self.search_field = self.driver.find_element_by_name("q")
        self.search_field.clear()

        # enter search keyword and submit
        self.search_field.send_keys("phones")
        self.search_field.submit()

        # get all the anchor elements which have product names
        # displayed currently on result page using
        # find_elements_by_xpath method
        products = self.driver\
            .find_elements_by_xpath
                ("//h2[@class='product-name']/a")

        # check count of products shown in results
        self.assertEqual(2, len(products))

    def tearDown(self):
        # close the browser window
        self.driver.quit()
```

```
if __name__ == '__main__':
    unittest.main(verbosity=2)
```

 应尽量避免在测试中隐式等待与显式等待混合使用，来处理同步问题。相比隐式等待，显式等待能提供更好的可操控性。

5.2 显式等待

显式等待是 WebDriver 中用于同步测试的另外一种等待机制。显式等待比隐式等待具备更好的操控性。与隐式等待不同，我们可以为脚本设置一些预置或定制化的条件，等待条件满足后再进行下一步测试。

显式等待可以只作用于仅有同步需求的测试用例。WebDriver 提供了 WebDriverWait 类和 expected_conditions 类来实现显式等待。

expected_conditions 类提供了一些预置条件，来作为测试脚本进行下一步测试的判断依据。让我们创建一个包含显式等待的简单的测试，条件是等待一个元素可见，代码如下。

```
from selenium import webdriver
from selenium.webdriver.common.by import By
from selenium.webdriver.support.ui import WebDriverWait
from selenium.webdriver.support import expected_conditions
import unittest

class ExplicitWaitTests(unittest.TestCase):
    def setUp(self):
        self.driver = webdriver.Firefox()
        self.driver.get("http://demo.magentocommerce.com/")

    def test_account_link(self):
        WebDriverWait(self.driver, 10)\
            .until(lambda s: s.find_element_by_id("select-
                language").get_attribute("length") == "3")
        account = WebDriverWait(self.driver, 10)\
```

```
            .until(expected_conditions.
                visibility_of_element_located
                    ((By.LINK_TEXT, "ACCOUNT")))
        account.click()

    def tearDown(self):
        self.driver.quit()

if __name__ == "__main__":
    unittest.main(verbosity=2)
```

在上面的测试中，显式等待条件是等到 Log In 链接在 DOM 中可见。

使用 visibility_of_element_located 方法来判断预期条件是否满足。该条件判断方法需要设置符合要求的定位策略和位置详细信息。脚本将一直查找目标元素是否可见，直到达到最长等待时间 10 秒。一旦根据指定的定位器找到了元素，预期条件判定方法将会把定位到的元素返回给测试脚本。

如果在设定的超时时间内，仍然没有通过定位器找到可见的目标元素，将会抛出 TimeoutException 异常。

5.3　expected_conditions 类

下表是 expected_conditions 类支持的在执行网页浏览器自动化操作时常常用到的一些通用的等待条件。

预期条件	描　述	参　数	实　例
element_to_be_clickable(locator)	等待通过定位器查找的元素可见并且可用，以便确定元素是可点击的。 此方法返回定位到的元素	locator： 一组（by,locator）	WebDriverWait(self.driver, 10).until(expected_conditions.element_to_be_clickable((By.NAME,"is_subscribed")))
element_to_be_selected(element)	等待直到指定的元素被选中	element： 是个 WebElement	subscription = self.driver.find_element_by_name("is_subscribed") WebDriverWait(self.driver, 10).until(expected_conditions.element_to_be_selected(subscription)

预期条件	描述	参数	实例
invisibility_of_element_located(locator)	等待一个元素在DOM中不可见或不存在	locator:一组（by,locator）	WebDriverWait(self.driver, 10).until(expected_conditions.invisibility_of_element_located((By.ID,"loading_banner")))
presence_of_all_elements_located(locator)	等待直到至少有一个定位器查找匹配到的目标元素出现在网页中。该方法返回定位到的一组WebElement	locator:一组（by,locator）	WebDriverWait(self.driver, 10).until(expected_conditions.presence_of_all_elements_located((By.CLASS_NAME,"input-text")))
presence_of_element_located(locator)	等待直到定位器查找匹配到的目标元素出现在网页中或可以在DOM中找到。该方法返回一个被定位到的元素	locator:一组（by,locator）	WebDriverWait(self.driver, 10).until(expected_conditions.presence_of_element_located((By.ID,"search")))
text_to_be_present_in_element(locator,text_)	等待直到元素能被定位到并且带有相应的文本信息	locator:一组（by,locator）text:需要被校验的文本内容	WebDriverWait(self.driver,10).until(expected_conditions.text_to_be_present_in_element((By.ID,"select-language"),"English"))
title_contains(title)	等待网页标题包含指定的大小写敏感的字符串。该方法在匹配成功时返回True，否则返回False	title:被校验的包含在标题中的字符串	WebDriverWait(self.driver, 10).until(expected_conditions.title_contains("Create New Customer Account"))
title_is(title)	等待网页标题与预期的标题相一致。该方法在匹配成功时返回True，否则返回False	title:网页的标题	WebDriverWait(self.driver, 10).until(expected_conditions.title_is("Create New CustomerAccount - MagentoCommerce Demo Store"))
visibility_of(element)	等待直到元素出现在DOM中，是可见的，并且宽和高都大于0。一旦其变成可见的，该方法将返回（同一个）WebElement	element:目标WebElement	first_name = self.driver.find_element_by_id("firstname") WebDriverWait(self.driver, 10).until(expected_conditions.visibility_of(first_name))
visibility_of_element_located(locator)	等待直到根据定位器查找的目标元素出现在DOM中，是可见的，并且宽和高都大于0。一旦其变成可见的，该方法将返回WebElement	locator:一组（by,locator）	WebDriverWait(self.driver, 10).until(expected_conditions.visibility_of_element_located((By.ID,"firstname")))

在下面的地址可看到预期条件判断的完整列表：selenium.googlecode /git/docs/api/py/webdriver_support/selenium.webdriver. support.expected_conditions.html#module-selenium.webdriver. support. expected_conditions。

在下面的章节中，让我们通过几个例子来了解更多的预期条件判断。

5.3.1 判断某个元素是否存在

正如在前面章节中看到的，expected_conditons 类提供了各种各样的预期等待条件，我们可以在脚本中实现。在下面的例子里，我们将等待一个元素变成可用或可点击。我们可以在 Ajax 应用较多的程序中使用这个预期等待条件，这样表单中一个字段是否可用取决于表单中别的字段或过滤器。该例子中，我们单击 Log In 链接，然后等待 Create an Account 按钮变成可点击的，这些元素都在登录页面上。最后我们单击 Create an Account 按钮，等待下一个页面加载完成并显示出来。

```python
def test_create_new_customer(self):
    # click on Log In link to open Login page
    self.driver.find_element_by_link_text("ACCOUNT").click()

    # wait for My Account link in Menu
    my_account = WebDriverWait(self.driver, 10)\
        .until(expected_conditions.visibility_of_element_located((By.
    LINK_TEXT, "My Account")))
    my_account.click()

    # get the Create Account button
    create_account_button = WebDriverWait(self.driver, 10)\
        .until(expected_conditions.element_to_be_clickable((By.LINK_
        TEXT, "CREATE AN ACCOUNT")))

    # click on Create Account button. This will displayed new account
    create_account_button.click()
    WebDriverWait(self.driver, 10)\
        .until(expected_conditions.title_contains("Create New Customer Account"))
```

我们等待并检查一个元素是否可用，可以用 element_to_be_clickable 预期条件。该方法需要指定定位策略或具体定位的位置。当目标元素变成可点击或者可用的时候，该方法返回定位到的目标元素给测试脚本。

前面的测试也介绍了通过检测标题是否含有指定的文本内容，来确定**创建新用户**页面是否加载成功。我们使用 title_contains 预期条件来检测,以确保指定的字符串能够与预期网页标题的子字符串相匹配。

5.3.2 判断是否存在 Alerts

我们也可以将显式等待应用于警告和页面框架中。例如,一个复杂的 JavaScript 处理过程或后端处理过程需要花费较多的时间把警告反馈给用户,这时可以用 alert_is_present 这个预期判断条件来实现,代码如下。

```
from selenium import webdriver
from selenium.webdriver.support.ui import WebDriverWait
from selenium.webdriver.common.by import By
from selenium.webdriver.support import expected_conditions
import unittest

class CompareProducts(unittest.TestCase):
    def setUp(self):
        self.driver = webdriver.Firefox()
        self.driver.get("http://demo.magentocommerce.com/")

    def test_compare_products_removal_alert(self):
        # get the search textbox
        search_field = self.driver.find_element_by_name("q")
        search_field.clear()

        # enter search keyword and submit
        search_field.send_keys("phones")
        search_field.submit()

        # click the Add to compare link
        self.driver.\
            find_element_by_link_text("Add to Compare").click()

        # wait for Clear All link to be visible
        clear_all_link = WebDriverWait(self.driver, 10)\
```

```python
            .until(expected_conditions.visibility_of_element_
            located((By.LINK_TEXT, "Clear All")))

        # click on Clear All link,
        # this will display an alert to the user
        clear_all_link.click()

        # wait for the alert to present
        alert = WebDriverWait(self.driver, 10)\
            .until(expected_conditions.alert_is_present())

        # get the text from alert
        alert_text = alert.text

        # check alert text
        self.assertEqual("Are you sure you would like
 to remove all products from your comparison?", alert_text)
        # click on Ok button
        alert.accept()

    def tearDown(self):
        self.driver.quit()

if __name__ == "__main__":
    unittest.main(verbosity=2)
```

上述的测试脚本,是验证从产品比较列表中移除所有的产品这个功能。当用户移除一个产品的时候,会收到是否确认的警告。Alert_is_present 预期判定条件就可以用来检测警告窗口是否出现,并且把警告窗口返回给脚本,以进行后续的动作。该脚本将会等待 10 秒的时间来检测警告窗口是否出现,如果没有出现就抛出异常。

5.4 预期条件判断的实践

正如在前面章节所了解到的,expected_conditions 类提供了多种定义好的预期等待条件。我们也可以通过 WebDriverWait 来自定义预期等待条件。当没有合适的预期等待条件可用的时候，自定义的预期等待条件也是非常有效的。

让我们来修改一个前面章节中创建好的测试脚本，实现一个自定义的预期条件判断，来检测下拉列表中可选项的数量。

```
def testLoginLink(self):
    WebDriverWait(self.driver, 10).until
      (lambda s: s.find_element_by_id
      ("select-language").get_attribute("length") == "3")

    login_link = WebDriverWait
      (self.driver, 10).until(expected_conditions.
      visibility_of_element_located((By.LINK_TEXT,"Log In")))
      login_link.click();
```

我们可以使用 Python 的 lambda 表达式，并且基于 WebDriverWait 来实现自定义的预期条件判断。上面的例子中，脚本将会等待 10 秒，直到 Select Language 下拉列表中有 8 个可选项。当下拉列表是通过 Ajax 调用来实现，并且脚本需要等待下拉列表中的所有选项都是可选择时，该预期条件判断是非常有用的。

5.5　章节回顾

在本章中，我们认识到元素等待机制对于构建高度稳定可靠的测试来说是必不可少的。我们学习了隐式等待，并且通过例子了解到了如何应用隐式等待作为通用的等待机制。显式等待可以提供更灵活的方式来同步进行测试。expected_conditions 类提供了多种内置的预期等待判定条件，我们在例子中也实践了一部分。

WebDriverWait 类提供了更加强大的自定义预期等待判定功能，超出了 expected_conditions。我们在下拉列表的例子中就实现了自定义的预期等待判定。

在下面的章节中，将会讲述如何通过使用 RemoteWebDriver 和 Selenium Server 使测试脚本在远程机器上执行，并且通过 Selenium Grid 实现脚本的并行执行，进而实现跨浏览器自动化测试。

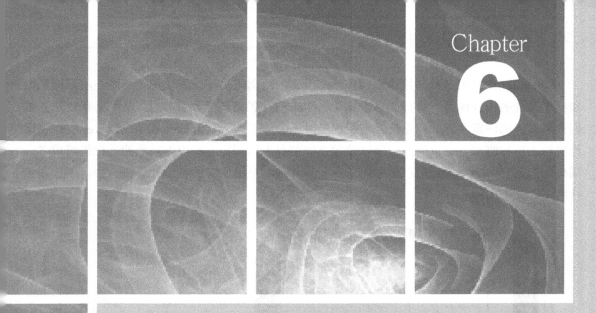

第 6 章
跨浏览器测试

第 6 章 跨浏览器测试

Selenium 支持由多种浏览器和操作系统组合的跨浏览器测试。该特性通过在不同浏览器和操作系统的组合场景下执行测试，来验证 Web 程序的跨浏览器兼容性，从而确保用户在他们喜好选择的浏览器和操作系统上使用程序时不会遇到问题。Selenium WebDriver 支持在远程机器上执行测试，并且能够把测试分发到安装有不同浏览器和操作系统的远程机器或者云端执行。到目前为止，我们已经学习了在安装各种浏览器驱动的本地计算机上如何创建和执行测试，如下图所示。

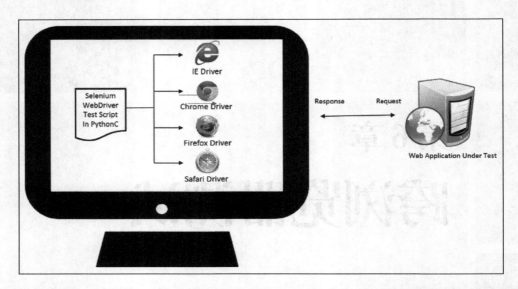

本章将学习如何在远程机器上执行测试，并且学习如何在由不同浏览器和操作系统组合成的分布式架构中的远程机器上批量执行跨浏览器测试。这种执行跨浏览器测试的实现方式将会节省大量的时间。

本章包含以下主题：

- Selenium Standalone Server 的下载和使用；
- 如何使用 Remote 类来实现在 Selenium Standalone Server 上执行测试；
- 在 Selenium Standalone Server 上执行测试；
- 为 Selenium Standalone Server 添加节点，从而为分布式执行创建一个 Grid；
- 在安装有多浏览器和操作系统组合的 Grid 上执行测试；
- 通过 Sauce Labs 和 BrowserStack 在云端执行测试。

6.1 Selenium Standalone Server

Selenium Standalone Server 是使 Selenium 具备在远程机器上执行测试能力的一个重要组件。我们需要通过使用 RemoteWebDriver 类来连接到 Selenium Standalone Server，从而实现在远程机器上执行测试。RemoteWebDriver 类通过特定的端口监听 Selenium 根据测试脚本所下达的命令。根据 RemoteWebDriver 类提供的配置选项，Selenium Server 可以选择启动的浏览器类型并且发送命令给浏览器。它几乎支持所有的浏览器，并且还可以基于 Appium 来实现对移动平台的支持。下面是 Selenium Server 在配置了不同类型浏览器的远程机器上执行测试的架构图。

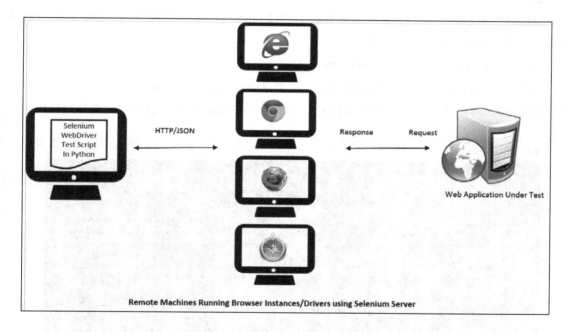

6.1.1 下载 Selenium Standalone Server

Selenium Standalone Server 是以 JAR 包的形式下载，可以从 seleniumhq 网站的 download 页面的 Selenium Server（原来的 Selenium RC Server）章节找到。写这本书时，Selenium Server 可下载版本是 2.41.0。你可以轻松地将 Selenium Standalone Server 的 JAR 包文件复制到远程机器上并启动服务。

Selenium Standalone Server 是用 Java 语言开发的，自我独立。在运行的时候，机器上需要安装 JRE(Java Runtime Environment)。在运行 Selenium Standalone Server 之前，要确保远程机器上已经安装了 JRE6 或者更高的版本。

6.1.2 启动 Selenium Standalone Server

Selenium Standalone Server 能以不同的模式或角色启动，在本章节我们采用的是 Standalone 模式启动。可以通过在远程机器上保存有 Selenium Standalone Server 的 JAR 包文件的目录下启动命令行，使用以下命令启动 Selenium Server，这是在 Windows 8 中启动 Selenium Standalone Server 的命令。

```
java -jar selenium-server-standalone-2.41.0.jar
```

Selenium Server 启动后，默认监听端口号是 4444)（http://<remote-machine-ip>:444）。在启动服务时，可以通过命令行更改端口号。下图是 Selenium Server 启动时的命令行输出。

Selenium Server 在远程机器上是以 HTTP Server 形式启动的，我们可以通过浏览器启动和查看该服务。在浏览器上输入 http://<remote-machine-ip>:4444/wd/hub/static/resource/hub.html，就可以看到如下图所示的服务启动后的界面。

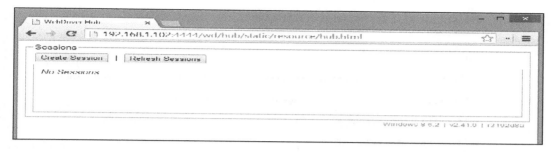

现在我们已启动并运行 Selenium Server,可以开始创建和执行测试了。

6.2 在 Selenium Standalone Server 上执行测试

要在 Selenium Server 上执行测试,我们需要使用 RemoteWebDriver。这个 Selenium Python 中的 Remote 类以客户端的身份与 Selenium Server 进行交互,从而实现在远程机器上运行测试。我们使用这个类来指示 Selenium Server 做出相应的操作,以在远程机器上运行测试,以及在指定的浏览器上运行测试命令。

除了 Remote 类之外,我们需要设置 desired_capabilities,即对浏览器和操作系统的配置,以及为了在 Selenium Standalone Server 上运行测试时要进行的一些其他配置。在此示例中,我们将指定运行测试的平台和浏览器名称,desired_capabilities 配置如下。

```
desired_caps = {}
desired_caps['platform'] = 'WINDOWS'
desired_caps['browserName'] = 'firefox'
```

接下来,将创建一个 Remote 类的实例并传递 desired_capabilities。当脚本执行时,该类将连接并请求 Selenium Server 启动 Windows 平台上的 Firefox 浏览器执行测试。

```
self.driver = webdriver.Remote('http://192.168.1.103:4444/wd/hub', desired_caps)
```

下面我们用 Remote 类代替 Firefox driver 实现一个之前创建的搜索测试。

```
import unittest
from selenium import webdriver

class SearchProducts(unittest.TestCase):
    def setUp(self):
        desired_caps = {}
        desired_caps['platform'] = 'WINDOWS'
```

```python
            desired_caps['browserName'] = 'firefox'

            self.driver = \
                webdriver.Remote('http://192.168.1.102:4444/wd/hub', desired_caps)
            self.driver.get('http://demo.magentocommerce.com/')
            self.driver.implicitly_wait(30)
            self.driver.maximize_window()

    def testSearchByCategory(self):
        # get the search textbox
        self.search_field = self.driver.find_element_by_name('q')
        self.search_field.clear()

        # enter search keyword and submit
        self.search_field.send_keys('phones')
        self.search_field.submit()

        # get all the anchor elements which have product names
        # displayed currently on result page using
        # find_elements_by_xpath method
        products = self.driver\
            .find_elements_by_xpath('//h2[@class=\'product-name\']/a')

        # check count of products shown in results
        self.assertEqual(2, len(products))

    def tearDown(self):
        # close the browser window
        self.driver.quit()

if __name__ == '__main__':
    unittest.main()
```

当执行此测试时，我们可以观察 Selenium Server 控制台的输出。它可以实时显示测试脚本与 Selenium Server 之间是如何进行交互的，以及已执行的命令和返回状态。下图是测试执行时控制台的信息。

在浏览器的 http:// <remote-machine-ip>:4444/wd/hub/static/resource/hub.html 页面可以发现，一个新的会话已创建。当鼠标停留在 Capabilities 链接上时，将显示用于该测试的 Capabilities 详细信息，如下图所示。

6.2.1 配置 IE 支持

Selenium Server 与 Firefox 绑定，默认支持 Firefox，但是要想在 Internet Explorer（IE）上执行测试，就需要在启动 Selenium Server 时指定 IE driver 可执行文件的路径。下面是在命令行中通过配置 webdriver.ie.driver 选项来指定 IE driver 可执行文件路径的命令。

```
java -Dwebdriver.ie.driver="C:\SeDrivers\IEDriverServer.exe" -jar
selenium-server-standalone-2.41.0.jar
```

通过指定 IE driver 路径启动 Selenium Server 后，就可以在远程机器的 IE 浏览器上执行测试了。

6.2.2 配置 Chrome 支持

与 IE driver 类似，要想在 Chrome 上执行测试，就需要指定 Chrome driver 可执行文件。下面是通过配置 webdriver.chrome.driver 选项指定 Chrome driver 可执行文件路径的命令。

```
java -Dwebdriver.ie.driver="C:\SeDrivers\IEDriverServer.exe" -Dwebdriver.
chrome.driver="C:\SeDrivers\chromedriver.exe" -jar selenium-server-standalone-2.41.0.jar
```

此时，Selenium Server 已同时支持在远程机器的 Internet Explorer 和 Chrome 浏览器上执行测试。

6.3 Selenium Grid

Selenium Grid 可以将测试分布在若干个物理或虚拟机器上，从而实现分布方式或并行方式执行测试。这样可以有效减少执行测试所需周期，同时实现跨浏览器测试来获得更快、更准确的结果反馈。我们可以使用云端现有的虚拟机建立 Grid。

Selenium Grid 能够在若干个节点或客户端上并行执行多个测试，这些节点或客户端都可以是不同浏览器和操作系统，从而支持混合的测试环境。Grid 使所有节点如下图展示的那样，在底层独立且透明地实现分布测试。

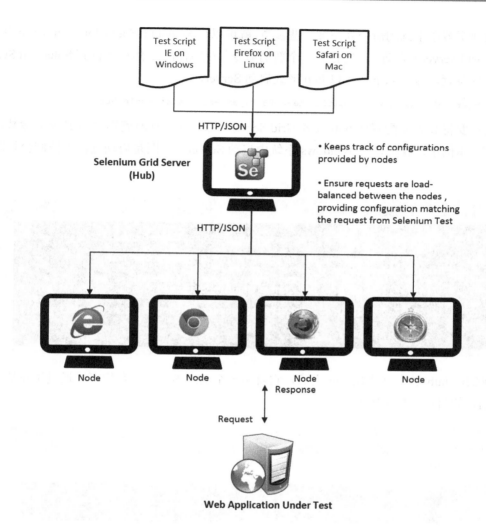

6.3.1 启动 hub

在分布式测试中，启动 Selenium Server 作为一个 hub，hub 提供所有可用的配置或属性信息给要执行的测试，然后 slave 机器（也称为节点）将连接到这个 hub。测试脚本运用 JSON Wire Protocol，并通过 Remote 类与 hub 交互来执行 Selenium 命令。更多关于 JSON Wire Protocol 的信息可以访问 google 网站。

hub 作为中心节点，接收测试命令并将它们分发给适当节点或符合匹配要求的节点。下面我们来将 Selenium Server 配置成 Grid，然后配置一些不同浏览器和操作系统组合的节点。

用前面章节中学到的命令并添加一些参数，就可以将 Selenium Standalone Server 作为 hub（也称为 Grid Server）启动。首先打开一个新的命令行/终端窗口，然后定位到 Selenium Server JAR 所在的位置。输入以下命令以 hub 形式启动 Server。

```
java -jar selenium-server-standalone-2.25.0.jar -port 4444 -role hub
```

注意：要将 Server 配置成 hub 或者 Grid Server，那么在启动时就需要指定 *-role* 参数，值为"hub"。本例中，我们是在 Windows 系统中启动 Server。下图是启动过程中控制台打印的信息。

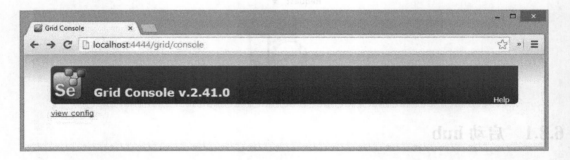

当我们以 hub 形式启动 Server 之后，它就是一个 Grid Server。我们可以通过浏览器查看 Grid 控制台的信息，如下图所示。

6.3.2 添加节点

现在我们已将 Selenium Server 作为 Grid Server 启动，接下来是在 Server 上添加并配置节点。

6.3.2.1 添加 IE 节点

先从安装有 Internet Explorer 的 Windows 节点开始添加。打开新的命令行或终端窗口并定位到 Selenium Server JAR 包文件所在的目录。输入以下命令启动节点并将它添加至 Grid。

```
java -Dwebdriver.ie.driver="C:\SeDrivers\IEDriverServer.exe" -jar
selenium-server-standalone-2.41.0.jar -role webdriver -browser
"browserName=internet explorer,version=10,maxinstance=1,platform=WINDOWS"
-hubHost 192.168.1.103 -port 5555
```

要将节点添加到 Grid，我们需要使用-role 参数并传递值"webdriver"，还需要使用-browser 参数配置浏览器信息。在这个例子中，通过 browserName 指定浏览器为 Internet Explorer，版本为 10，maxinstance 为 1，平台为 Windows。其中 maxinstance 值是告诉 Grid 可以支持多少并发浏览器实例。

要将节点连接到 hub 或 Grid Server，我们还需要指定-hubHost 参数与 Grid Server 的主机名或 IP 地址。最后，指定节点就可以连接到 hub 上对应的端口。

当我们执行上述命令后，节点将被启动，Grid 控制台上会出现如下图所示的配置。

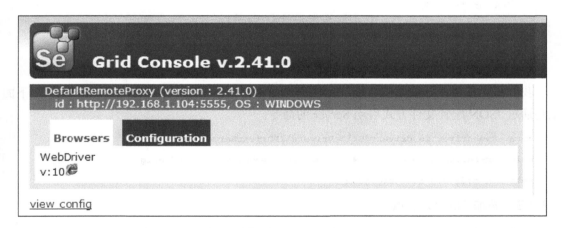

除了上面的方法外，也可以通过 JSON 格式的配置文件来添加节点。JSON 配置文件代码如下。

```
{
  "class": "org.openqa.grid.common.RegistrationRequest",
  "capabilities": [
    {
```

```
    "seleniumProtocol": "WebDriver",
    "browserName": "internet explorer",
    "version": "10",
    "maxInstances": 1,
    "platform" : "WINDOWS"
    }
 ],
 "configuration": {
  "port": 5555,
  "register": true,
  "host": "192.168.1.103",
  "proxy": "org.openqa.grid.selenium.proxy. DefaultRemoteProxy",
  "maxSession": 2,
  "hubHost": "192.168.1.100",
  "role": "webdriver",
  "registerCycle": 5000,
  "hub": "http://192.168.1.100:4444/grid/register",
  "hubPort": 4444,
  "remoteHost": "http://192.168.1.102:5555"
 }
}
```

我们可以在命令行参数中传递 JSON 配置文件 "selenium-node-win-ie10.cfg.json"，下面就是通过 JSON 配置文件方式启动 Server 的命令。

```
java -Dwebdriver.ie.driver="C:\SeDrivers\IEDriverServer.exe" -jar
selenium-server-standalone-2.41.0.jar -role webdriver -nodeConfig
selenium-node-win-ie10.cfg.json
```

6.3.2.2 添加 Firefox 节点

现在开始添加 Firefox 节点。打开新的命令行或终端窗口，定位到 Selenium Server JAR 包文件所在目录下，输入以下命令启动节点并添加至 Grid。

```
java -jar selenium-server-standalone-2.41.0.jar -role webdriver -browser
"browserName=firefox,version=27,maxinstance=2,platform=WINDOWS" -hubHost
localhost -port 6666
```

在此，我们设置 maxinstance 值为 2，也就是告诉 Grid 可以同时支持两个 Firefox 实例。

Firefox 节点启动后，Grid 控制台就会出现如下图所示的配置。

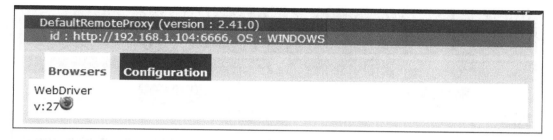

6.3.2.3　添加 Chrome 节点

接下来添加 Chrome 节点。打开新的命令行或终端窗口，定位到 Selenium Server JAR 包文件所在目录下，输入以下命令启动节点并添加至 Grid。

```
java -Dwebdriver.chrome.driver="C:\SeDrivers\chromedriver.exe" -jar
selenium-server-standalone-2.41.0.jar -role webdriver -browser
"browserName=chrome,version=35,maxinstance=2,platform=WINDOWS" -hubHost localhost -port
7777
```

下图是 Chrome 节点启动后 Grid 控制台出现的配置信息。

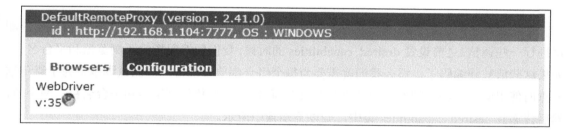

6.4　Mac OS X 的 Safari 节点

我们已经从 Windows 系统添加了 IE、Firefox 和 Chrome 节点，现在我们要从 Mac 系统添加一个 Safari 节点。打开一个新的终端窗口，定位到 Selenium Server JAR 包文件所在目录下，输入以下命令启动节点并添加至 Grid。

```
java -jar selenium-server-standalone-2.41.0.jar -role webdriver -browser
"browserName=safari,version=7,maxinstance=1,platform=MAC" -hubHost 192.168.1.104 -port
8888
```

下图是该节点启动后 Grid 控制台上出现的所有配置信息。

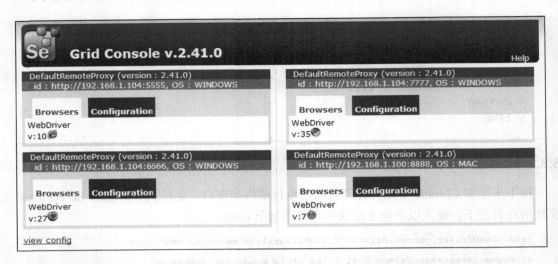

现在，我们配置好了 Selenium Grid，让我们尝试在这个 Grid 上执行测试吧。

6.5 在 Grid 上执行测试

在配置有不同浏览器和操作系统组合的 Grid 上执行测试之前，需要对我们之前创建的测试进行一些调整。之前设置 desired_capabilities 的时候，我们硬编码指定了浏览器和平台名称。如果这些值被硬编码了，那么我们就需要为每个组合都单独准备一个测试脚本。为了避免这一点而使用同一个测试脚本来测试所有的组合，我们需要参数化浏览器和平台名称这两个值，然后传递到 desired_capabilities 类中，如以下步骤中所示。

（1）从命令行中将浏览器和平台的值传递给测试脚本。例如，如果要在 Windows 和 Chrome 组合上执行测试，可以通过以下方式在命令行中运行脚本。

```
python grid_test.py WINDOWS chrome
```

（2）如果要在 Mac 和 Safari 组合上执行测试，使用以下命令运行脚本。

```
python grid_test.py MAC safari
```

（3）要实现这一点，需要在以下的测试类中添加 PLATFORM 和 BROWSER 两个全局属性，并分别设置一个默认值，以防从命令行运行脚本时没有提供值。

```
class SearchProducts(unittest.TestCase):
    PLATFORM = 'WINDOWS'
```

```
BROWSER = 'firefox'
```

（4）接下来，在setUp()方法中参数化这些desired_capabilities，如下面的代码所示。

```
desired_caps = {}
desired_caps['platform'] = self.PLATFORM
desired_caps['browserName'] = self.BROWSER
```

（5）最后，读取命令行参数值并传递给脚本，从而为PLATFORM和BROWSER属性赋值。

```
if __name__ == '__main__':
    if len(sys.argv) > 1:
        SearchProducts.BROWSER = sys.argv.pop()
        SearchProducts.PLATFORM = sys.argv.pop()
    unittest.main()
```

（6）就是这样。现在我们的测试已经准备好处理任何给定的环境组合了。以下是完整代码。

```
import sys
import unittest
from selenium import webdriver

class SearchProducts(unittest.TestCase):
    PLATFORM = 'WINDOWS'
    BROWSER = 'firefox'

    def setUp(self):
        desired_caps = {}
        desired_caps['platform'] = self.PLATFORM
        desired_caps['browserName'] = self.BROWSER

        self.driver = \
            webdriver.Remote('http://192.168.1.104:4444/wd/hub', desired_caps)
        self.driver.get('http://demo.magentocommerce.com/')
        self.driver.implicitly_wait(30)
        self.driver.maximize_window()
```

```python
    def testSearchByCategory(self):
        # get the search textbox
        self.search_field = self.driver.find_element_by_name('q')
        self.search_field.clear()

        # enter search keyword and submit
        self.search_field.send_keys('phones')
        self.search_field.submit()

        # get all the anchor elements which have product names
        # displayed currently on result page using
        # find_elements_by_xpath method
        products = self.driver.\
            find_elements_by_xpath('//h2[@class=\'product-name\']/a')

        # check count of products shown in results
        self.assertEqual(2, len(products))

    def tearDown(self):
        # close the browser window
        self.driver.quit()

if __name__ == '__main__':
    if len(sys.argv) > 1:
        SearchProducts.BROWSER = sys.argv.pop()
        SearchProducts.PLATFORM = sys.argv.pop()
    unittest.main(verbosity=2)
```

（7）打开新的命令行或终端窗口，定位至脚本所在位置的目录，输入以下命令执行测试。你将看到 Grid 会自动连接到与给定平台和浏览器匹配的节点并在该节点上执行测试。

```
python grid_test.py MAC safari
```

6.6 在云端执行测试

为了实现跨浏览器测试，我们在之前的步骤中搭建了本地 Grid。搭建本地 Grid 需要给物理或虚拟机器配置不同浏览器和操作系统。可是获取这些硬件和软件设备是需要很大的成本和努力的，而且你还需要在这些设备的更新和补丁等方面投入大量的精力，这不是每个人或团队都能负担得起的。

但现在你可以轻松地将虚拟测试 lab 外包给第三方云测试提供商，而不必花费精力投资和搭建跨浏览器测试 lab。Sauce Labs 和 BrowserStack 是领先的基于云端的跨浏览器测试云提供商。这两个都支持超过 400 种不同的浏览器和操作系统配置，包括移动和平板设备，并支持在他们的云端环境中运行 Selenium WebDriver 测试。

在本节中，我们将在 Sauce Labs 中安装并执行测试。用 BrowserStack 执行测试的步骤也是类似的。

下面我们使用 Sauce Labs 搭建并执行测试，步骤如下。

（1）首先需要一个免费的 Sauce Labs 账号。可以从 Sauce Labs 官网注册免费账号，获取用户名和访问密钥。Sauce Labs 在云端环境中提供了执行测试所需的所有硬件和软件等基础配置。

（2）登录后从 Sauce Labs 主页获取访问密钥，如下图所示。

（3）修改之前 Grid 中运行时所创建的测试，并添加步骤使其可以运行在 Sauce Labs 上。

（4）在测试脚本中添加 Sauce 用户名和访问密钥，更改 Grid 地址为 Sauce Grid 地址并传递该用户名和访问密钥，如下面代码所示。

```
import sys
import unittest
```

```python
from selenium import webdriver

class SearchProducts(unittest.TestCase):
    PLATFORM = 'WINDOWS'
    BROWSER = 'phantomjs'
    SAUCE_USERNAME = 'upgundecha'
    SUACE_KEY = 'c6e7132c-ae27-4217-b6fa-3cf7df0a7281'

    def setUp(self):
        desired_caps = {}
        desired_caps['platform'] = self.PLATFORM
        desired_caps['browserName'] = self.BROWSER

        sauce_string = self.SAUCE_USERNAME + ':' + self.SUACE_KEY

        self.driver = \
            webdriver.Remote('http://' + sauce_string +
            '@ondemand.saucelabs.com:80/wd/hub',   desired_caps)
        self.driver.get('http://demo.magentocommerce.com/')
        self.driver.implicitly_wait(30)
        self.driver.maximize_window()

    def testSearchByCategory(self):
        # get the search textbox
        self.search_field = self.driver.find_element_by_name('q')
        self.search_field.clear()

        # enter search keyword and submit
        self.search_field.send_keys('phones')
        self.search_field.submit()

        # get all the anchor elements which have product names
        # displayed currently on result page using
        # find_elements_by_xpath method
```

```python
        products = self.driver.\
            find_elements_by_xpath('//h2[@class=\'product-name\']/a')

        # check count of products shown in results
        self.assertEqual(2, len(products))

    def tearDown(self):
        # close the browser window
        self.driver.quit()

if __name__ == '__main__':
    if len(sys.argv) > 1:
        SearchProducts.BROWSER = sys.argv.pop()
        SearchProducts.PLATFORM = sys.argv.pop()
    unittest.main(verbosity=2)
```

（5）打开命令行或终端窗口，进入脚本所在目录，输入以下命令执行测试。

```
python sauce_test.py "OS X 10.9" "Safari"
```

 可以在 saucelabs 网站获取 Sauce Labs 支持的平台列表。

测试执行时，它将连接到 Sauce Labs 的 Grid Server 并请求所需的操作系统和浏览器配置。Sauce 为我们的测试分配虚拟机并在给定的配置上运行。

（6）我们可以在 Sauce Dashboard 上监视执行状态，如下图所示。

Session	Environment	Tags	Build	Results	End ▼	Run Time
unnamed job	10.9 7			Running View Fullscreen | Spy		

我们还可以在 Sauce 会话部分进一步了解测试执行过程中究竟发生了什么。Sauce 会话页面显示了测试执行的细节，包括 Selenium 命令、截图、Selenium 日志以及执行过程的截屏，如下图所示。

 也可以用 Sauce Connect 在内部服务器上更安全地测试你的应用程序。Sauce Connect 会在你的机器和 Sauce 云中创建一个安全的通道。

6.7 章节回顾

在本章中,我们学习了如何使用 Selenium Standalone Server 在远程机器上执行测试。Selenium Standalone Server 能够在远程机器上对应用程序进行任意浏览器和操作系统组合的跨浏览器测试。这不但增加了测试覆盖率,而且能确保应用程序在期望的组合上执行。

然后,我们搭建了 Selenium Grid,并在分布式架构中执行测试。Selenium Grid 可以在对多个机器上透明地执行,从而降低跨浏览器测试的复杂性,也减少了测试执行时间。

我们还考虑了基于云端的跨浏览器测试提供商,在 Sauce Labs 上执行测试。Sauce Labs 提供了执行测试所需的所有基础配置,支持上百种不同的组合,成本更低。

在下一章中,我们将学习如何使用 Appium 和 Selenium WebDriver 测试移动端应用程序,在此过程中会用到一些本章中学习到的概念。Appium 支持在 iOS 和 Android 系统上测试原生、混合以及移动 Web 应用程序。我们将展示用 Appium 针对移动端测试的示例。

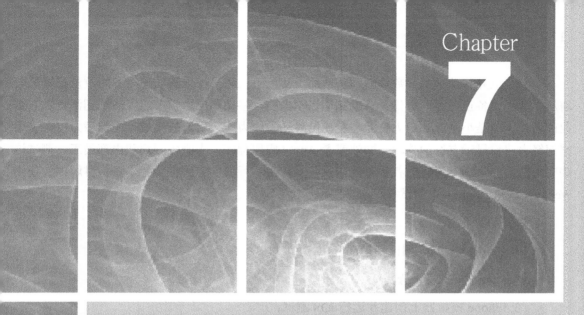

第 7 章
移动端测试

第 7 章 移动端测试

随着全世界移动端用户数量的不断增加，智能手机和平板电脑的使用者的数量已经显著增加。智能设备正在逐渐取代台式机和笔记本电脑，移动端应用程序已经渗透到消费者与企业市场。无论是小企业还是大企业，在使用移动端作为连接用户的渠道方面都潜力巨大。大家都努力建设显示友好的移动端网站或者 App（应用程序）来服务客户和内部员工。测试这些应用能否在市场上流行的各种移动端设备上正常运行变得至关重要。本章将讲解如何使用 WebDriver 和 Appium 来测试移动端应用程序。

本章包含以下主题：

- 如何使用 Appium 测试移动端应用；
- Appium 的安装和配置；
- 在 iPhone 模拟器上创建并运行 iOS 测试；
- 在真机上创建并运行 Android 测试。

7.1 认识 Appium

Appium 是一个开源的自动化测试框架，可以用来测试基于 iOS、Android 和 Firefox OS 平台的原生与混合的应用。该框架使用 Selenium WebDriver，在执行测试时用于和 Selenium Server 通信的是 JSON Wire Protocol。在 Selenium 2 中，Appium 将取代 iPhoneDriver 和 AndroidDriver API，并用于测试移动互联网应用程序。

Appium 允许我们使用，甚至扩展现有的 Selenium WebDriver 框架来构建测试脚本。由于 Appium 是通过 Selenium WebDriver 来驱动测试脚本的，因此只要有对应的 Selenium client library 存在，就可以使用相应的语言来创建测试脚本。下图是 Appium 对不同平台和应用类型的支持情况的覆盖地图。

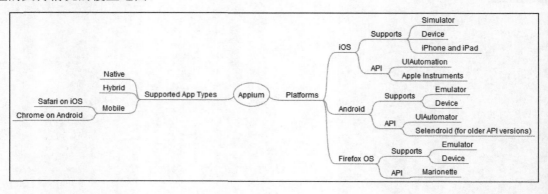

7.1.1 Appium 支持的应用类型

Appium 支持以下应用类型的测试。

- **原生应用**。原生应用是指适用于特定平台的，即使用该平台所支持的语言和框架来构建的。例如，iPhone 和 iPad 上面的应用都是使用 Objective-C 和 iOS SDK 来开发的；同样，Android 应用是使用 Java 和 Android SDK 来开发的。在程序运行的时候，原生应用会更加流畅和稳定。它们是使用原生框架来构建用户交互界面的。
- **移动端 Web 应用**。移动端 Web 应用是服务端应用，是使用 PHP、Java 或者 ASP.NET 这样的服务端技术来构建的，并且使用 jQuery Mobile、Sencha Touch 等一些技术渲染用户页面以模拟本地 UI。
- **混合应用**。类似于原生应用程序，混合应用是运行在移动设备上并且通过一些互联网技术（HTML5、CSS 和 JavaScript）来实现的。混合应用使用移动设备的浏览器引擎来渲染 HTML 页面，并且通过使用 WebView 在本地容器中处理 JavaScript 脚本。这种处理方式可以使混合应用具备访问一些移动 Web 应用不能访问的设备（比如相机、加速计、传感器和本地存储器）的能力。

7.1.2 Appium 环境准备

在开始学习更多关于 Appium 的知识之前，首先需要了解一些基于 iOS 和 Android 平台的工具。

Appium 是基于 Node.js 实现的，在 Mac OS X 和 Windows 平台上都有对应的 Standalone GUI 的 Node.js 包。我们可以使用 Mac OS X 平台上内置在 Node.js 框架内的 Appium Standalone GUI。

7.1.2.1 安装 Xcode

我们需要在 Mac OS X 系统上使用 Xcode 4.6.3 或者更高的版本，来测试 iOS 平台下的应用。

写本书时，使用的是 Xcode 5.1 版本。可以在 App Store 或者苹果开发者网站下载。

安装完 Xcode 后，从应用程序菜单启动它，然后单击 Preferences | Downloads。安装 Command Line Tools 和其他的 iOS SDK，用来测试在不同版本 iOS 平台下的应用程序，如下图所示。

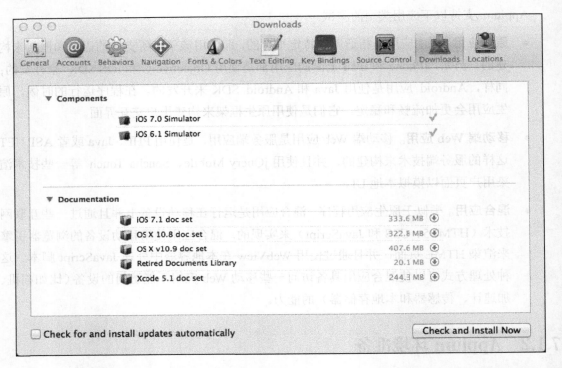

为了在真机上运行测试脚本，需要首先安装好 Provision Profile，并且打开 USB 调试模式。

启动 iPhone 模拟器并验证其是否正常工作。可以通过菜单 Xcode | Open Developer Tool | iOS Simulator 来启动模拟器。在模拟器中启动 Safari 浏览器，并且打开如下图所示 Web 应用的样例。

7.1.2.2　安装 Android SDK

我们需要安装 Android SDK 来测试 Android 应用。从 developer.android 网站可以获取最新版本的 Android SDK。安装完成以后，确保 ANDROID_HOME 已经成功添加到环境变量的 path 中。

完整安装步骤参考：developer.android/sdk/installing/index.html?pkg=tools。

有关最新的 Appium 安装要求和细节,请访问相关网站了解。

7.1.2.3 安装 Appium Python client

在完成此书时,Appium Python client 是完全符合 Selenium 3.0 规范草案的。这有助于更加方便地使用 Python 和 Appium 来编写移动端测试脚本。可以通过以下命令来安装 Appium Python client。

```
pip install Appium-Python-Client
```

可以访问 pypi.python 网站查看更多关于 Appium Python client 安装包的信息。

7.2 安装 Appium

在使用 Appium 测试移动应用之前，我们首先需要下载和安装 Appium。我们选择安装 Appium GUI 版。如果希望在 iPhone 或者 iPad 上运行 iOS 测试，那么就需要在装有 Mac OS X 系统的机器上安装 Appium。如果是测试 Android 应用程序，需要在装有 Windows 或者 Linux 系统的机器上安装 Appium。在 Mac OS X 系统上安装 Appium 是非常简单的。可以从 appium 网站下载最新版本的 Appium 安装文件。

具体安装步骤如下。

（1）在网站首页单击 **Download Appium** 按钮，即可直接跳转到下载页面，如下图所示。

（2）从列表中选择你正在使用的操作系统对应的安装版本，如下图所示。

 在下面的例子中，我们将在 Mac OS X 系统上使用 Appium。

（3）在 Mac 系统中，可以通过运行安装程序来安装 Appium，并且把 Appium 复制到 Applications 目录下。

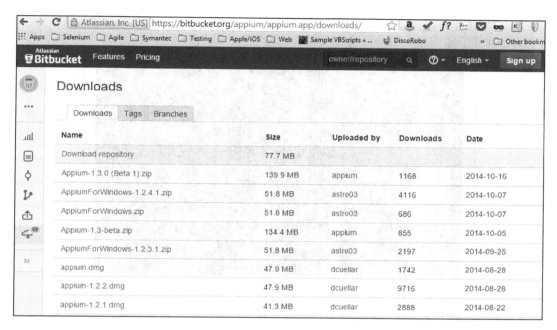

当第一次从 Applications 菜单启动 Appium 的时候，它将会要求授权来运行 iOS 模拟器。

 默认情况下，Appium 启动后的 URL 和端口是 http://127.0.0.1:4723 或 localhost。该 URL 正是测试脚本发指令对应的地址。我们将基于 iPhone 的 Safari 浏览器来测试本书中的样例程序的移动版本。

（4）在如下图所示的 Appium 的主界面，单击 Apple 图标，打开 iOS 设置对话框。

（5）在 iOS 设置对话框中，选中 Force Device 复选框，并且在下拉列表中选择 iPhone 型号。另外，选中 Use Mobile Safari 复选框，如下图所示。

（6）在 Appium 主界面，单击 Launch 按钮启动 Appium Server。

Inspector 元素定位器：Appium 的元素定位工具叫 Appium Inspector，可以通过在 Appium 主界面单击放大镜的图标来启动它。

该定位器提供了定位被测应用的多种分析方式。其中主要的特性就是告诉我们这些 UI 元素在移动应用程序中是如何被使用的，包括它们的层次结构以及元素的属性，通过这些我们可以设置元素的定位器。它也可以模拟应用程序上的各种操作手势，并看到在模拟器上执行的效果。它还能够记录下用户在移动应用上的操作步骤。

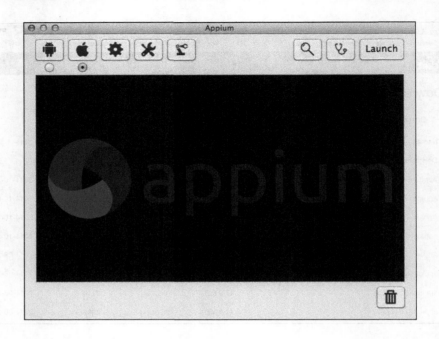

7.3 iOS 测试

Appium 使用多种自带自动化测试框架来驱动测试执行,并且支持基于 Selenium WebDriver JSON 无线协议的 API 调用。对于 iOS 应用程序的自动化测试,Appium 是利用 Apple 组件之一的 UI Automation 特性来实现的。

Appium 作为一个 HTTP Server 来接收测试脚本通过 JSON Wire Protocol 传输过来的指令。Appium 发送这些指令给 Apple 组件。这样,这些指令就可以在启动了被测应用的模拟器或者真机上执行测试。这个过程中,Appium 把 JSON 指令翻译成 UI Automation 能够理解的 JavaScript 命令。UI Automaiton 负责在模拟器或者真机上启动和关闭应用程序。这个过程如下图所示。

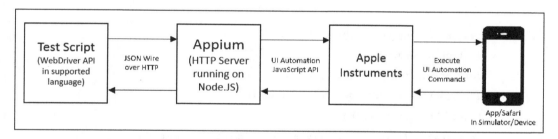

当在模拟器或者真机的应用程序上执行测试指令的时候,被测应用程序会发送响应给 UI Automation,然后以 JavaScript 的响应格式返回给 Appium。Appium 再把 UI Automation JavaScript 响应翻译成符合 Selenium WebDriver JSON Wire Protocol 的响应信息,并把它返回给测试脚本。

现在,我们已经成功启动了 Appium,接下来创建一个基于 iPhone Safari 浏览器,检查 Web 应用搜索功能的测试用例。创建一个新的测试类 SearchProductsOnIPhone,代码如下。

```python
import unittest
from appium import webdriver

class SearchProductsOnIPhone(unittest.TestCase):
    def setUp(self):
        desired_caps = {}
        # platform
```

```python
        desired_caps['device'] = 'iPhone Simulator'
        # platform version
        desired_caps['version'] = '7.1'
        # mobile browser
        desired_caps['app'] = 'safari'

        # to connect to Appium server use RemoteWebDriver
        # and pass desired capabilities
        self.driver = \
            webdriver.Remote("http://127.0.0.1:4723/wd/hub", desired_caps)
        self.driver.get("http://demo.magentocommerce.com/")
        self.driver.implicitly_wait(30)
        self.driver.maximize_window()

    def test_search_by_category(self):
        # click on search icon
        self.driver.find_element_by_xpath
        ("//a[@href='#header-search']").click()
        # get the search textbox
        self.search_field = self.driver.find_element_by_name("q")
        self.search_field.clear()
        # enter search keyword and submit
        self.search_field.send_keys("phones")
        self.search_field.submit()

        # get all the anchor elements which have product names
        # displayed currently on result page using
        # find_elements_by_xpath method
        products = self.driver\
            .find_elements_by_xpath
            ("//div[@class='category-products']/ul/li")

        # check count of products shown in results
        self.assertEqual(2, len(products))
```

```python
    def tearDown(self):
        # close the browser window
        self.driver.quit()

if __name__ == '__main__':
    unittest.main(verbosity=2)
```

需要使用 RemoteWebDriver 来调用 Appium 执行测试用例。为了在期望的平台上使用 Appium，需要执行以下代码。

```python
desired_caps = {}
# platform
desired_caps['device'] = 'iPhone Simulator'
# platform version
desired_caps['version'] = '7.1'
# mobile browser
desired_caps['app'] = 'safari'
```

desired_caps['device'] 命令是用 Appium 来指定在哪个平台上运行测试脚本。上面的例子中，是使用 iPhone 模拟器。如果要在 iPad 上执行测试，那么就要指定 iPad 模拟器。

当在真机上运行测试脚本时，需要为 desired_caps['device']赋值为 iPhone 或者 iPad。Appium 会选择通过 USB 连接到 Mac 电脑上的设备来执行脚本。

desired_caps['version']命令用来设置 iPhone/iPad 模拟器的版本。上面的例子中，使用的是 iOS 7.1 版本的模拟器，该版本在写本书的时候是 iOS 的最新版本。

desired_caps['app']用来设置要启动的目标应用。上面的例子中，启动的是 Safari 浏览器。

最后，需要通过 RemoteWebDriver 连接到 Appium Server，并且设置好相应的配置。下面是创建一个 Remote 实例的代码。

```python
self.driver = webdriver.Remote
    ("http://127.0.0.1:4723/wd/hub", desired_caps)
```

后面的测试脚本用来调用 Selenium API 实现与应用的交互操作。运行测试过程中，将会看到 Appium 通过测试脚本建立会话，并在 iPhone 模拟器上启动 Safari 应用程序。Appium 将会在模拟器窗口中一步步地执行对于 Safari 应用的测试。

7.4 Android 测试

Appium 对于 Android 应用程序的测试是通过调用内置在 Android SDK 中的 UI Automator 来实现的。过程非常类似于基于 iOS 平台的测试。

Appium 作为一个 HTTP Server 来接收测试脚本通过 JSON Wire Protocol 传输过来的指令。Appium 发送这些指令给 UI Automator。这样，这些指令就可以在启动了被测应用的模拟器或者真机上执行测试。这个过程中，Appium 把 JSON 指令翻译成 UI Automator 能够理解的 Java 命令。这个过程如下图所示。

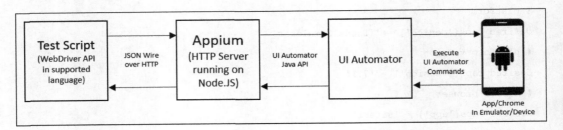

当在模拟器或者真机的应用程序上执行测试指令的时候，被测应用程序会发送响应给 UI Automator，然后返回给 Appium。Appium 再把 UI Automator 响应翻译成符合 Selenium WebDriver JSON Wire Protocol 的响应信息，并把它返回给测试脚本。

在 Android 平台上测试一个应用程序和在 iOS 平台上的操作非常相似。一般在 Android 平台，我们会使用真机来取代模拟器（与 iOS 平台的 simulator 不同，在 Android 社区被称为 emulator）。下面将使用同样的程序来测试 Android 平台下的 Chrome 浏览器。

在本次测试中，将使用 Samsung Galaxy S III 型号的手机。测试之前首先要安装好 Chrome 浏览器。Chrome 浏览器可以从应用商店下载。然后需要把该手机连接到运行有 Appium 的电脑上。

下面就要在 Android 上开始测试工作了。我们是在 Android 真机上执行测试脚本的，首先要确保手机已经成功安装了 Chrome 浏览器并且成功连接到电脑上了。运行下面的命令可以获取连接到电脑上的虚拟机和真机的列表。

./adb devices

Android Debug Bridge(adb) 是内置在 Android SDK 中的命令行工具，通过它可以与模拟器实例或者真机进行通信。

上面的命令可以获取到已经连接到主机上的 Android 设备列表。如下图所示，可以看到在上面的例子中已经建立连接的真机。

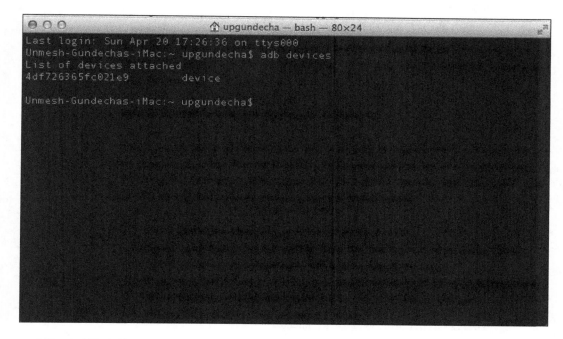

我们可以修改前面为 iOS 测试写过的脚本，使其能够在 Android 平台运行。创建一个新的测试类 SearchProductsOnAndroid。复制下面的代码到新创建的测试类中。

```
import unittest
from appium import webdriver

class SearchProductsOnAndroid(unittest.TestCase):
    def setUp(self):
        desired_caps = {}
        # platform
        desired_caps['device'] = 'Android'
        # platform version
        desired_caps['version'] = '4.3'
        # mobile browser
        desired_caps['app'] = 'Chrome'
```

```python
        # to connect to Appium server use RemoteWebDriver
        # and pass desired capabilities
        self.driver = \
            webdriver.Remote("http://127.0.0.1:4723/wd/hub", desired_caps)
        self.driver.get("http://demo.magentocommerce.com/")
        self.driver.implicitly_wait(30)

    def test_search_by_category(self):

        # click on search icon
        self.driver.find_element_by_xpath
            ("//a[@href='#header-search']").click()
        # get the search textbox
        self.search_field = self.driver.find_element_by_name("q")
        self.search_field.clear()

        # enter search keyword and submit
        self.search_field.send_keys("phones")
        self.search_field.submit()

        # get all the anchor elements which have product names
        # displayed currently on result page using
        # find_elements_by_xpath method
        products = self.driver\
            .find_elements_by_xpath
                ("//div[@class='category-products']/ul/li")

        # check count of products shown in results
        self.assertEqual(2, len(products))

    def tearDown(self):
        # close the browser window
        self.driver.quit()
```

```
if __name__ == '__main__':
    unittest.main(verbosity=2)
```

在该实例中，我们给 desired_caps['device']指定的值是 Android，表明将会在 Android 平台上运行测试。

接下来，可以看到 desired_caps['version']设置的是 Android 4.3 版本（Jelly Bean）。需要在 Android 系统的 Chrome 浏览器上执行测试，因此 desired_caps['app']设置的值是 Chrome。

Appium 将会调用通过 adb 返回的设备列表中的第一个来运行测试，并通过前面提到的配置，来实现在目标设备上启动 Chrome 浏览器，然后开始执行测试脚本，如下图所示。

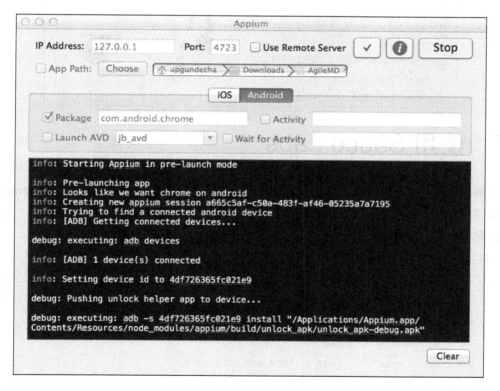

下面是在真机上运行测试脚本时的截图。

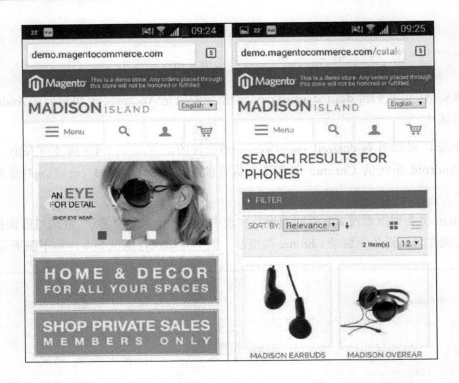

7.5 使用 Sauce Labs

在第 6 章讲到跨浏览器测试时，已经提到了 Sauce Labs。Sauce 也支持通过 Appium 来测试移动端应用程序。事实上，Appium 本身就是由 Sauce Labs 来开发和支持的项目。通过非常小的配置修改，就可以在 Sauce Labs 上运行移动端的测试脚本，代码如下。

```
import unittest
from appium import webdriver

class SearchProductsOnIPhone(unittest.TestCase):
    SAUCE_USERNAME = 'upgundecha'
    SUACE_KEY = 'c6e7132c-ae27-4217-b6fa-3cf7df0a7281'

    def setUp(self):
        desired_caps = {}
        desired_caps['browserName'] = "Safari"
        desired_caps['platformVersion'] = "7.1"
```

```python
        desired_caps['platformName'] = "iOS"
        desired_caps['deviceName'] = "iPhone Simulator"

        sauce_string = self.SAUCE_USERNAME + ':' + self.SUACE_KEY

        self.driver = \
            webdriver.Remote('http://' + sauce_string +
            '@ondemand.saucelabs.com:80/wd/hub', desired_caps)
        self.driver.get('http://demo.magentocommerce.com/')
        self.driver.implicitly_wait(30)
        self.driver.maximize_window()

    def test_search_by_category(self):
        # click on search icon
        self.driver.find_element_by_xpath("//a[@href=
        '#header-search']").click()
        # get the search textbox
        self.search_field = self.driver.find_element_by_name("q")
        self.search_field.clear()

        # enter search keyword and submit
        self.search_field.send_keys("phones")
        self.search_field.submit()

        # get all the anchor elements which have
        # product names displayed
        # currently on result page using
        # find_elements_by_xpath method
        products = self.driver\
            .find_elements_by_xpath
            ("//div[@class='category-products']/ul/li")

        # check count of products shown in results
        self.assertEqual(2, len(products))
```

```
    def tearDown(self):
        # close the browser window
        self.driver.quit()

if __name__ == '__main__':
    unittest.main(verbosity=2)
```

当执行完移动端测试以后,可以在 Sauce Labs 的面板上看到运行的结果和录像回放。Sauce Labs 上有很多现成的 SDK 的组合和设置,在 Sauce Labs 执行测试可以节省大量用在本地环境配置 Appium 的时间和精力。

7.6 章节回顾

本章我们学习了如果在移动设备上测试应用程序。我们了解了 Appium,它已经成为 Selenium 用来测试移动端应用程序的一个核心特性。我们完成了 Appium 的安装和配置,并且通过它测试了本书的样例程序的移动端版本。

我们分别在 iPhone 模拟器和 Android 设备上测试移动互联网程序。通过使用 Appium,我们可以使用任何一种有 WebDriver 类库的编程语言来测试各类移动端应用程序。

在下面章节中,将会讲解一些良好的编码实践,比如通过 Selenium WebDriver 来实现 Page Object 模式和数据驱动测试。

第 8 章
Page Object 与数据驱动测试

第 8 章 Page Object 与数据驱动测试

本章将介绍两类重要的设计模式，这些设计模式有助于提升我们自动化测试框架的可扩展性与可维护性。我们将一起学习如何用数据驱动的模式结合 Python 库去构建 Selenium 测试脚本。

在本章的第二部分，我们还将学习用 Page Object 的模式创建高可维护与健壮性的测试脚本。将元素定位器和底层调用从测试脚本中分离出来形成抽象层，如同实现应用程序的各个功能（就像用户在浏览器窗口中体验到的内容一样）。

本章包含以下主题：

- 什么是数据驱动测试；
- 如何用数据驱动的模式（ddt）库结合 unittest 库构建数据驱动测试；
- 什么是 Page Object 模式以及如何使用该模式创建维护性好的测试；
- 结合测试样例实现一个 Page Object 模式的测试。

8.1 数据驱动测试

通过使用数据驱动测试的方法，我们可以在需要验证多组数据的测试场景中，使用外部数据源实现对输入值与期望值的参数化，从而避免在测试中仅使用硬编码的数据。

这种方法对于测试步骤相同而使用不同的"输入值与期望值"的测试场景尤其适用。下表列举了一个用于验证登录场景的数据组合。

场景描述	测试数据	预期结果
有效的用户名和密码	一组有效的用户名和密码	用户正常登录并且收到成功提示信息
无效的用户名和密码	一组无效的用户名和密码	用户收到登录失败的信息
有效的用户名与无效的密码	一个有效的用户名与一个无效的密码	用户收到登录失败的信息

我们只需创建一个测试脚本就可以处理上表的测试数据和条件的各个组合。

使用数据驱动的模式，根据业务逻辑分解测试数据，并且定义变量，使用外部的 CSV 或表格里的数据使其参数化，从而避免使用原测试脚本中固定的数据。这种方式可以将测试脚本与测试数据分离，使得测试脚本在不同数据集合下高度复用。

数据驱动的模式不仅可以帮助我们增加类似复杂条件场景的测试覆盖，还可以极大地减少我们对测试代码的编写和维护工作量。

接下来的部分，我们将改用 Python ddt 库，以数据驱动的模式创建前面章节已经实现过的测试。

8.2 使用 ddt 执行数据驱动测试

ddt 的库可以将测试中的变量进行参数化。例如，我们可以通过定义一个数组来实现数据驱动测试。

ddt 的库包含一组类和方法用于实现数据驱动测试。

8.2.1 安装 ddt

可以使用下面的命令来下载与安装 ddt。

```
pip install ddt
```

更多有关 ddt 的信息可以访问 pypi.python 网站。

8.2.2 设计一个简单的数据驱动测试

我们将一个之前使用数据硬编码的搜索场景测试，转换成用数据驱动模式进行测试，并且使得脚本可以搜索多种类别的商品。

为了创建数据驱动测试，我们需要在测试类上使用@ddt 装饰符，在测试方法上使用@data 装饰符。@data 装饰符把参数当作测试数据，参数可以是单个值、列表、元组、字典。对于列表，需要用@unpack 装饰符把元组和列表解析成多个参数。

接下来实现这个搜索测试，传入搜索关键词和期望结果，代码如下。

```
import unittest
from ddt import ddt, data, unpack
from selenium import webdriver


@ddt
class SearchDDT(unittest.TestCase):
    def setUp(self):
```

```python
        # create a new Firefox session
        self.driver = webdriver.Firefox()
        self.driver.implicitly_wait(30)
        self.driver.maximize_window()

        # navigate to the application home page
        self.driver.get("http://demo.magentocommerce.com/")

    # specify test data using @data decorator
    @data(("phones", 2), ("music", 5))
    @unpack
    def test_search(self, search_value, expected_count):
        # get the search textbox
        self.search_field = self.driver.find_element_by_name("q")
        self.search_field.clear()
        # enter search keyword and submit.
        # use search_value argument to pass data
        self.search_field.send_keys(search_value)
        self.search_field.submit()

        # get all the anchor elements which have
        # product names displayed
        # currently on result page using
        # find_elements_by_xpath method
        products = self.driver.find_elements_by_xpath
            ("//h2[@class='product-name']/a")

        # check count of products shown in results
        self.assertEqual(expected_count, len(products))

    def tearDown(self):
        # close the browser window
        self.driver.quit()
```

```python
if __name__ == '__main__':
    unittest.main(verbosity=2)
```

在上面的代码里,我们在@data装饰符中把元组列表作为参数,紧接着在@unpack装饰符中把元组解析成多个参数。在test_search()方法中,search_value与expected_count两个参数用来接收元组解析的数据。

```python
# specify test data using @data decorator
    @data(("phones", 2), ("music", 5))
    @unpack
    def test_search(self, search_value, expected_count):
```

当我们运行测试脚本的时候,ddt把测试数据转换为有效的Pyhotn标识符,生成名称更有意义的测试方法。例如上面的测试,ddt将生成如下图展示的方法名。

```
Microsoft Windows [Version 6.2.9200]
(c) 2012 Microsoft Corporation. All rights reserved.

C:\Users\amitr>cd C:\Users\amitr\Desktop\Mrunmayee\Final\setests_final

C:\Users\amitr\Desktop\Mrunmayee\Final\setests_final>python SearchDDT.py
test_search_1___phones___2_ (__main__.SearchDDT) ... ok
test_search_2___music___5_ (__main__.SearchDDT) ... ok

----------------------------------------------------------------------
Ran 2 tests in 70.885s

OK

C:\Users\amitr\Desktop\Mrunmayee\Final\setests_final>
```

8.3 使用外部数据的数据驱动测试

在先前的例子里,我们在脚本中直接提供了测试数据。然而,有时你会发现所需要的测试数据在测试脚本外部已经存在了,诸如一个文本文件、电子表格或是数据库。这可以使得我们的测试脚本与测试数据分离开来,可以方便我们每次更新与维护测试脚本,而不用担心测试数据。

下面我们一起学习如何借助外部的CSV(逗号分隔值)文件或是Excle表格数据来实现ddt。

8.3.1 通过CSV获取数据

结合前面的测试脚本,我们在@data装饰符中使用解析的外部的CSV(testdata.csv)来换掉之前的测试数据。其中 CSV 数据文件如下图所示。

```
testdata.csv - Notepad
File Edit Format View Help
CategoryOrProduct,NumberOfProducts
phones,2
music,5
iphone 5s,0
```

接下来，我们在@data 装饰符中实现 get_data()方法，其中包括路径、CSV 文件名。这个方法调用 CSV 库去读取文件并且返回一行数据。

```python
import csv, unittest
from ddt import ddt, data, unpack
from selenium import webdriver


def get_data(file_name):
    # create an empty list to store rows
    rows = []
    # open the CSV file
    data_file = open(file_name, "rb")
    # create a CSV Reader from CSV file
    reader = csv.reader(data_file)
    # skip the headers
    next(reader, None)
    # add rows from reader to list
    for row in reader:
        rows.append(row)
    return rows


@ddt
class SearchCsvDDT(unittest.TestCase):
    def setUp(self):
        # create a new Firefox session
```

```python
        self.driver = webdriver.Firefox()
        self.driver.implicitly_wait(30)
        self.driver.maximize_window()

        # navigate to the application home page
        self.driver.get("http://demo.magentocommerce.com/")

    # get the data from specified csv file by
    # calling the get_data function
    @data(*get_data("testdata.csv"))
    @unpack
    def test_search(self, search_value, expected_count):
        self.search_field =self.driver.find_element_by_name("q")
        self.search_field.clear()

        # enter search keyword and submit.
        self.search_field.send_keys(search_value)
        self.search_field.submit()

        # get all the anchor elements which have
        # product names displayed
        # currently on result page using
        # find_elements_by_xpath method
        products = self.driver.find_elements_by_xpath
            ("//h2[@class='product-name']/a")
        expected_count = int(expected_count)
        if expected_count > 0:
            # check count of products shown in results
            self.assertEqual(expected_count, len(products))
        else:
            msg = self.driver.find_element_by_class_name ("note-msg")
            self.assertEqual ("Your search returns no results.", msg.text)

    def tearDown(self):
```

```
        # close the browser window
        self.driver.quit()

if __name__ == '__main__':
    unittest.main()
```

测试执行时，@data 将调用 get_data()方法去读取外部数据文件，并将数据逐行返回给 @data。

8.3.2 通过 Excel 获取数据

用 Excel 来维护测试数据是最常用的做法。这还可以帮助非技术人员很轻松地添加一行需要的测试数据。结合上面的例子，我们把数据整理到 Excel 中，如下图所示。

A	B
Category/Product	NumberOfProducts
phones	2
music	5
iphone 5s	0

读取 Excel 文件，我们需要用到另外一个叫 xlrd 的库，其安装命令如下。

pip install xlrd

> xlrd 库提供了读取工作簿、工作表以及单元格的方法。如果需要往表格中写数据，则需要用到 xlwt 库。另外，openpyxl 提供了对电子表格可读可写的功能。

接下来我们修改 get_data()方法，试着从外部的电子表格获取测试数据，代码如下。

```
import xlrd, unittest
from ddt import ddt, data, unpack
from selenium import webdriver


def get_data(file_name):
    # create an empty list to store rows
    rows = []
```

```python
    # open the specified Excel spreadsheet as workbook
    book = xlrd.open_workbook(file_name)
    # get the first sheet
    sheet = book.sheet_by_index(0)
    # iterate through the sheet and get data from rows in list
    for row_idx in range(1, sheet.nrows):
        rows.append(list(sheet.row_values(row_idx, 0, sheet.ncols)))
    return rows


@ddt
class SearchExcelDDT(unittest.TestCase):
    def setUp(self):
        # create a new Firefox session
        self.driver = webdriver.Firefox()
        self.driver.implicitly_wait(30)
        self.driver.maximize_window()

        # navigate to the application home page
        self.driver.get("http://demo.magentocommerce.com/")


    # get the data from specified Excel spreadsheet
    # by calling the get_data function
    @data(*get_data("TestData.xlsx"))
    @unpack
    def test_search(self, search_value, expected_count):
        self.search_field = self.driver.find_element_by_name("q")
        self.search_field.clear()

        # enter search keyword and submit.
        self.search_field.send_keys(search_value)
        self.search_field.submit()

        # get all the anchor elements which have
        # product names displayed
```

```
        # currently on result page using
        # find_elements_by_xpath method
        products = self.driver.find_elements_by_xpath
            ("//h2[@class='product-name']/a")
        if expected_count > 0:
            # check count of products shown in results
            self.assertEqual(expected_count, len(products))
        else:
            msg = self.driver. find_element_by_class_name("note-msg")
            self.assertEqual("Your search returns no results.", msg.text)

    def tearDown(self):
        # close the browser window
        self.driver.quit()

if __name__ == '__main__':
    unittest.main()
```

类似读取 CSV 文件一样，@data 将调用 get_data()方法去读取外部 Excel 文件，并将数据逐行返回给@data。

通过数据库获取数据

如果你想要从数据库的库表中获取数据，那么你同样需要修改 get_data() 方法，并且通过 DB 相关的库来连接数据库、SQL 查询来获取测试数据。

8.4 Page Object 设计模式

到目前为止，我们已经掌握了在 Python unittest 单元测试框架中，编写 Selenium WebDriver 测试脚本的方法，并且可以在测试类中根据测试用例的步骤使用合适的定位器。然后，随着陆续越来越多的测试场景加入自动化测试用例，那么与之对应的测试脚本就变得越来越难以维护，甚至代码变得很脆弱。

开发可维护性和可重用的测试脚本，对于持续自动化测试是很重要的。其重要性完全不亚于我们被测产品的代码标准。

如何去解决这些问题呢？如果你是一个开发工程师，你可能会调动很多原则和方法，诸如 Don't Repeat Yourself（DRY，高级码农的信条，不做重复的自己，不写重复的代码）或定期代码重构的方法。Page Object 模式，是使用 Selenium 的广大同行最为公认的一种设计模式。在设计测试时，把元素和方法按照页面抽象出来，分离成一定的对象，然后再进行组织。它相较之前的 Facade 模式（为系统中的一组接口所提供的一个一致的界面）又更进了一步。

Page Object 模式，创建一个对象来对应页面的一个应用。因此，我们可以为每个页面定义一个类，并为每个页面的属性和操作构建模型。这就相当于在测试脚本和被测的页面功能中分离出一层，屏蔽了定位器、底层处理元素的方法和业务逻辑，取而代之的是，Page Object 会提供一系列的 API 来处理页面功能。

测试应该在更上层使用这些页面对象，在底层页面中的属性或操作的任何更改都不会中断测试。Page Object 模式具有以下几个优点。

- 抽象出对象可以最大程度地降低开发人员修改页面代码对测试的影响，所以，你仅需要对页面对象进行调整，而对测试没有影响；
- 可以在多个测试用例中复用一部分测试代码；
- 测试代码变得更易读、灵活、可维护。

接下来，我们一起将前面章节已经实现过的测试脚本进行重构，对原应用程序页面抽象

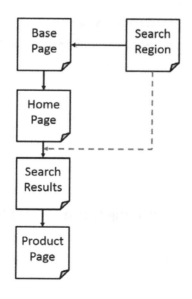

出对象。在这个示例中，我们将针对被测页面创建如下图所示的结构。首先我们要实现一个 Base Page 对象（可以理解为其他页面所要用到的模板），对应的 Base 对象提供了其他页面需要用到的功能区块。例如，所有的页面都用到搜索功能，那么我们可以基于 Base Page 创建一个 Search 对象，接下来我们分别为首页、搜索结果页以及产品页面各自创建类。

8.4.1 测试准备

在我们开始用 Page Object 模式设计测试之前，首先得先实现一个名为 BaseTestCase 的类，用于给我们提供 setUp() 和 tearDown() 两种方法，以便后续我们写每个类都可以拿来复用。创建名为 basetestcase.py 的脚本，代码细节如下。

```python
import unittest
from selenium import webdriver


class BaseTestCase(unittest.TestCase):
    def setUp(self):
        # create a new Firefox session
        self.driver = webdriver.Firefox()
        self.driver.implicitly_wait(30)
        self.driver.maximize_window()

        # navigate to the application home page
        self.driver.get('http://demo.magentocommerce.com/')

    def tearDown(self):
        # close the browser window
        self.driver.quit()
```

8.4.2 BasePage 对象

BasePage 对象相当于所有页面对象中的父对象，同时可以提供公共部分的代码。创建名为 base.py 的脚本，代码细节如下。

```python
from abc import abstractmethod
```

```python
class BasePage(object):
    """ All page objects inherit from this """

    def __init__(self, driver):
        self._validate_page(driver)
        self.driver = driver

    @abstractmethod
    def _validate_page(self, driver):
        return

    """ Regions define functionality available through all page objects """
    @property
    def search(self):
        from search import SearchRegion
        return SearchRegion(self.driver)

class InvalidPageException(Exception):
    """ Throw this exception when you don't find the correct page """
    pass
```

我们增加了一个名为_validate_page()的抽象方法,继承 BasePage 的 page 对象将实现这个方法,目的是在能够使用属性和操作之前,验证页面是否已经加载到浏览器。

另外,我们还创建了 search 属性用于返回 SearchRegion 对象。类似于一个页面对象,SearchRegion 相当于每个页面都用到的搜索框。所以接下来其他页面对象都可以共享这个 BasePage 类。

最后,我们也实现了_validate_ page()方法中用到的 InvalidPageException,如果页面验证失败,InvalidPageException 将会被抛出。

8.4.3 实现 Page Object

现在,我们可以为每个页面实现 Page Object 了。

(1) 首先,我们定义 HomePage,创建 homepage.py,实现 HomePage 类,代码如下。

```python
from base import BasePage
from base import InvalidPageException


class HomePage(BasePage):

    _home_page_slideshow_locator = 'div.slideshow-container'

    def __init__(self, driver):
        super(HomePage, self).__init__(driver)

    def _validate_page(self, driver):
        try:
            driver.find_element_by_class_name (self._home_page_slideshow_locator)
        except:
            raise InvalidPageException ("Home Page not loaded")
```

我们要遵循的一点是将定位器字符串与它们的使用位置分离开。我们可以创建一个"_"前缀的私有变量，例如，用_home_page_slideshow_locator 变量来保存应用程序首页的 slideshow 组件的定位器字符串。我们可以利用这个来确认浏览器是否正常加载了首页。

```python
_home_page_slideshow_locator = 'div.slideshow-container'
```

然后，我们可以创建_validate_page()方法，通过判断 slideshow 元素是否已经显示在首页上了，来判断首页是否被加载。

（2）接下来，我们实现 SearchRegion 类，包括 searchFor()方法，该方法用于返回 SearchResults 类对应的搜索结果页面。创建一个新的脚本 search.py，并且实现这两个类，代码如下。

```python
from base import BasePage
from base import InvalidPageException
from product import ProductPage


class SearchRegion(BasePage):
    _search_box_locator = 'q'

    def __init__(self, driver):
        super(SearchRegion, self).__init__(driver)
```

```python
    def searchFor(self, term):
        self.search_field = self.driver.find_element_by_name(self._search_box_locator)
        self.search_field.clear()
        self.search_field.send_keys(term)
        self.search_field.submit()
        return SearchResults(self.driver)

class SearchResults(BasePage):
    _product_list_locator = 'ul.products-grid > li'
    _product_name_locator = 'h2.product-name a'
    _product_image_link = 'a.product-image'
    _page_title_locator = 'div.page-title'

    _products_count = 0
    _products = {}

    def __init__(self, driver):
        super(SearchResults, self).__init__(driver)
        results = self.driver.find_elements_by_css_selector(self._product_list_locator)
        for product in results:
            name = product.find_element_by_css_selector(self._product_name_locator).text
            self._products[name] = product.find_element_by_css_selector(self._product_image_link)

def _validate_page(self, driver):
    if 'Search results for' not in driver.title:
        raise InvalidPageException('Search results not loaded')


@property
def product_count(self):
    return len(self._products)


def get_products(self):
    return self._products
```

```python
def open_product_page(self, product_name):
    self._products[product_name].click()
    return ProductPage(self.driver)
```

(3)最后,我们要实现 ProductPage 类,这个类包括了很多有关商品的一些属性。访问一个商品的详细页面,可以通过 SearchResults 类,打开搜索结果中一个具体的产品。创建 **product.py** 脚本文件实现 **ProductPage** 类,代码如下。

```python
from base import BasePage
from base import InvalidPageException


class ProductPage(BasePage):
    _product_view_locator          = 'div.product-view'
    _product_name_locator          = 'div.product-name span'
    _product_description_locator   = 'div.tab-content div.std'
    _product_stock_status_locator  = 'p.availability span.value'
    _product_price_locator         = 'span.price'

    def __init__(self, driver):
        super(ProductPage, self).__init__(driver)

    @property
    def name(self):
        return self.driver.\
            find_element_by_css_selector\
              (self._product_name_locator)\
            .text.strip()

    @property
    def description(self):
        return self.driver.\
            find_element_by_css_selector\
              (self._product_description_locator)\
            .text.strip()
```

```
    @property
    def stock_status(self):
        return self.driver.\
            find_element_by_css_selector
                (self._product_stock_status_locator)\
            .text.strip()


    @property
    def price(self):
        return self.driver.\
            find_element_by_css_selector
                (self._product_price_locator)\
            .text.strip()


    def _validate_page(self, driver):
        try:
            driver.find_element_by_css_selector (self._product_view_locator)
        except:
            raise InvalidPageException ('Product page not loaded')
```

你还可以进一步添加一些测试场景，例如在商品页点击商品添加到购物车，或者比较商品，又或者通过商品属性返回相关的商品。

8.4.4 构建 Page Object 模式测试实例

结合之前的准备，我们可以构建完整测试了。下面我们创建一个用于检验应用程序搜索功能的测试，使用 BaseTestCase 类并调用我们之前创建的页面对象。该测试首先创建一个 HomePage 实例，并调用 searchFor()方法返回 SearchResults 实例。然后调用 SearchResults 类中的 open_product_page()方法，来打开返回的搜索结果中商品的详情页，进而检查商品属性。

相关脚本 searchtest.py，以及其中 SearchProductTest 测试类代码如下。

```
import unittest
from homepage import HomePage
from BaseTestCase import BaseTestCase
```

```python
class SearchProductTest(BaseTestCase):
    def testSearchForProduct(self):
        homepage = HomePage(self.driver)
        search_results = homepage.search.searchFor('earphones')
        self.assertEqual(2, search_results.product_count)
        product = search_results.open_product_page('MADISON EARBUDS')
        self.assertEqual('MADISON EARBUDS', product.name)
        self.assertEqual('$35.00', product.price)
        self.assertEqual('IN STOCK', product.stock_status)

if __name__ == '__main__':
    unittest.main(verbosity=2)
```

注意，在这里我们并没有写 setUp() 和 tearDown() 两个方法。我们只需要继承 BaseTestCase 类，直接使用其已经实现过的方法即可。如果你有特殊的测试场景，当然也可以重载这两个方法。

通过对上述例子完整的学习，我们已经掌握了一个页面完整工作流的 Page Object 设计测试的实践。你也可以通过类似的模式对如购物车、账号注册、登录等场景设计测试了。

8.5 章节回顾

本章我们认识了编写数据驱动的测试方法，以及构建可复用、易量化、维护性好的 Page Object 模式测试脚本。其中数据驱动的方法，可以将我们的测试数据与测试脚本分离开来，使得我们可以使用更复杂的测试数据，而不用编写新的测试脚本。结合 unittest 和 ddt 库，我们可以轻松实现多种外部数据源的测试数据获取。另外，关于如何用 Page Object 模式设计测试脚本，以及如何对一个简单的业务场景设计一个高可维护性的测试脚本做了详细的介绍。

在接下来的章节，我们将学习 Selenium WebDriver API 的一些高级特性，例如常用的截屏、屏幕录制、模拟鼠标与键盘操作、操作 cookies 等。

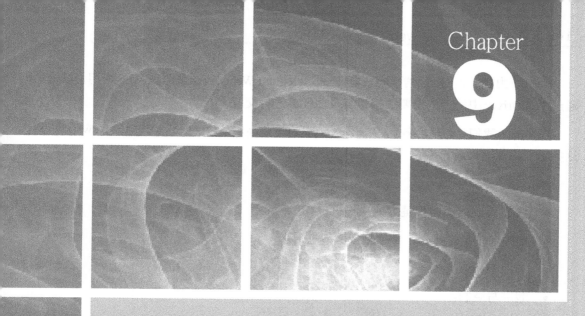

第 9 章
Selenium WebDriver 的高级特性

第 9 章 Selenium WebDriver 的高级特性

到目前为止，我们已经学习了如何使用 Selenium WebDriver 来测试 Web 应用，以及如何通过 WebDriver 中的一些主要的接口与页面元素进行交互。

在本章，我们将进一步探讨 WebDriver 中的一些高级 API，当测试较复杂的应用场景时，这些功能将派上用场。

本章包含以下主题：

- 通过 Action 类模拟键盘或鼠标事件；
- 模拟一些鼠标操作，例如拖拽、双击等；
- 调用 JavaScript；
- 截屏与录制；
- 处理导航与 cookies；
- 键盘与鼠标事件。

9.1 键盘与鼠标事件

WebDriver 高级应用的 API，允许我们模拟简单到复杂的键盘和鼠标事件，如拖拽操作、快捷键组合、长按以及鼠标右键操作。这些都是通过使用 WebDriver 的 Python API 中 ActionChains 类实现的。

下表列出 ActionChains 类中一些关于键盘和鼠标事件的重要方法。

方　法	描　述	参　数	样　例
click(on_element=None)	单击元素操作	on_element: 指被单击的元素。如果该参数为 None，将单击当前鼠标位置	click(main_link)
click_and_hold(on_element=None)	对元素按住鼠标左键	on_element: 指被单击且按住鼠标左键的元素。如果该参数为 None，将单击当前鼠标位置	click_and_hold(gmail_link)
double_click(on_element=None)	双击元素操作	on_element: 指被双击的元素。如果该参数为 None，将双击当前鼠标位置	double_click(info_box)

续表

方 法	描 述	参 数	样 例
drag_and_drop(source, target)	鼠标拖动	source: 鼠标拖动的源元素。 target: 鼠标释放的目标元素	drag_and_drop(img, canvas)
key_down(value, element=None)	仅按下某个键,而不释放。这个方法用于修饰键(如 Ctrl、Alt 与 Shift 键)	key:指修饰键。Key 的值在 Keys 类中定义。 target:按键触发的目标元素,如果为 None,则按键在当前鼠标聚焦的元素上触发	key_down(Keys.SHIFT)\\ send_keys('n')\\ key_up(Keys.SHIFT)
key_up(value, element=None)	用于释放修饰键	key:指修饰键。Key 的值在 Keys 类中定义。 target:按键触发的目标元素,如果为 None,则按键在当前鼠标聚焦的元素上触发	
move_to_element(to_element)	将鼠标移动至指定元素的中央	to_element: 指定的元素	move_to_element(gmail_link)
perform()	提交(重放)已保存的动作		perform()
release(on_element=None)	释放鼠标	on_element: 被鼠标释放的元素	release(banner_img)
send_keys(keys_to_send)	对当前焦点元素的键盘操作	keys_to_send: 键盘的输入值	send_keys("hello")
send_keys_to_element(element, keys_to_send)	对指定元素的键盘操作	element: 指定的元素。 keys_to_send: 键盘的输入值	send_keys_to_element(firstName, "John")

获取更多细节可访问 selenium.googlecode/git/docs/api/py/webdriver/selenium.webdriver.common.action_chains.html。

Interactions API 目前不支持 Safari 浏览器,同时在其他浏览器上也有一定的局限性。更多详情请访问 code.google/p/selenium/wiki/Advanced UserInteractions。

9.1.1 键盘事件

接下来我们创建一个测试脚本，用来模拟一个组合键的操作。在这个简单的场景中，当我们按下 Shift+N 组合键时，label 标签会改变颜色。代码如下。

```python
from selenium import webdriver
from selenium.webdriver.common.by import By
from selenium.webdriver.support.ui import WebDriverWait
from selenium.webdriver.support import expected_conditions
from selenium.webdriver.common.action_chains import ActionChains
from selenium.webdriver.common.keys import Keys
import unittest

class HotkeyTest(unittest.TestCase):
    URL = "rawgit/jeresig/jquery.hotkeys/master/test-static-05.html"

    def setUp(self):
        self.driver = webdriver.Chrome()
        self.driver.get(self.URL)
        self.driver.implicitly_wait(30)
        self.driver.maximize_window()

    def test_hotkey(self):
        driver = self.driver

        shift_n_label = WebDriverWait(self.driver, 10).\
            until(expected_conditions.visibility_of_element_
            located((By.ID, "_shift_n")))

        ActionChains(driver).\
            key_down(Keys.SHIFT).\
            send_keys('n').\
            key_up(Keys.SHIFT).perform()
        self.assertEqual("rgba(12, 162, 255, 1)",
```

```
                            shift_n_label.value_of_css_
                            property("background-color"))

    def tearDown(self):
        self.driver.close()
if __name__ == "__main__":
    unittest.main(verbosity=2)
```

通过使用 ActionChains 类，我们可以实现组合键操作。在上面的示例中，我们联合 key_down()、send_key()与 key_up()三个方法模拟真人操作 Shift+N 组合键。

```
ActionChains(driver).\
    key_down(Keys.SHIFT).\
    send_keys('n').\
    key_up(Keys.SHIFT).perform()
```

当调用 ActionChains 类的方法时，它不会立即执行，而是会将所有的操作按顺序存放在一个队列里，当调用 perform()方法时，队列中的事件会依次执行。

9.1.2 鼠标事件

下面演示一个调用 ActionChains 类中的 move_to_element()方法实现鼠标移动的示例。这个方法类似于 onMouseOver 事件。move_to_element()方法是将光标从当前位置移动到指定的元素。

```
from selenium import webdriver
from selenium.webdriver.common.by import By
from selenium.webdriver.support.ui import WebDriverWait
from selenium.webdriver.support import expected_conditions
from selenium.webdriver.common.action_chains import ActionChains
import unittest


class ToolTipTest (unittest.TestCase):
    def setUp(self):
        self.driver = webdriver.Firefox()
```

```python
        self.driver.get("//jqueryui/tooltip/")
        self.driver.implicitly_wait(30)
        self.driver.maximize_window()

    def test_tool_tip(self):
        driver = self.driver

        frame_elm = driver.find_element_by_class_name("demo-frame")
        driver.switch_to.frame(frame_elm)

        age_field = driver.find_element_by_id("age")
        ActionChains(self.driver).move_to_element(age_field).perform()

        tool_tip_elm = WebDriverWait(self.driver, 10)\
            .until(expected_conditions.visibility_of_element_
            located((By.CLASS_NAME, "ui-tooltip-content")))

        # verify tooltip message
        self.assertEqual("We ask for your age only for statistical
        purposes.", tool_tip_elm.text)

    def tearDown(self):
        self.driver.close()

if __name__ == "__main__":
    unittest.main(verbosity=2)
```

9.1.2.1 双击操作

调用 ActionChains 类中的 double_click()方法实现鼠标对元素的双击操作,代码如下。

```
from selenium import webdriver

from selenium.webdriver.common.action_chains import ActionChains
import unittest

class DoubleClickTest (unittest.TestCase):
```

```python
    URL = "api.jquery/dblclick/"

    def setUp(self):
        self.driver = webdriver.Chrome()
        self.driver.get(self.URL)
        self.driver.maximize_window()

    def test_double_click(self):
        driver = self.driver
        frame = driver.find_element_by_tag_name("iframe")
        driver.switch_to.frame(frame)
        box = driver.find_element_by_tag_name("div")

        # verify color is Blue
        self.assertEqual("rgba(0, 0, 255, 1)",
                         box.value_of_css_property("background-color"))

        ActionChains(driver).move_to_element\
            ( driver.find_element_by_tag_name("span"))\.perform()

        ActionChains(driver).double_click(box).perform()

        # verify Color is Yellow
        self.assertEqual("rgba(255, 255, 0, 1)",
                         box.value_of_css_property("background-color"))

    def tearDown(self):
        self.driver.close()

if __name__ == "__main__":
    unittest.main(verbosity=2)
```

9.1.2.2 鼠标拖动

调用 ActionChains 类中的 drag_and_drop() 方法实现鼠标的拖放操作。这个方法拖动源元素，然后在目标元素的位置释放源元素。示例如下。

```python
from selenium import webdriver
from selenium.webdriver.common.action_chains import ActionChains
import unittest


class DragAndDropTest (unittest.TestCase):

    URL = "jqueryui/resources/
           demos/droppable/default.html"

    def setUp(self):
        self.driver = webdriver.Firefox()
        self.driver.get(self.URL)
        self.driver.maximize_window(30)
        self.driver.maximize_window()

    def test_drag_and_drop(self):
        driver = self.driver

        source = driver.find_element_by_id("draggable")
        target = driver.find_element_by_id("droppable")

        ActionChains(self.driver).drag_and_drop(source, target).perform()
        self.assertEqual("Dropped!", target.text)

    def tearDown(self):
        self.driver.close()

if __name__ == "__main__":
    unittest.main(verbosity=2)
```

9.2 调用 JavaScript

在执行某些特殊操作或测试 JavaScript 代码时，WebDriver 还提供了调用 JavaScript 的方法。WebDriver 类包含的相关方法见下表。

方法	描述	参数	示例
execute_async_ script(script, *args)	异步执行 JS 代码	script:被执行的 JS 代码。args:JS 代码中的任意参数	driver.execute_async_script("return document.title")
execute_script(script, *args)	同步执行 JS 代码	script:被执行的 JS 代码。args:JS 代码中的任意参数	driver.execute_ script("return document.title")

接下来创建的测试用到了工具方法,该工具方法在使用 JavaScript 方法对元素执行操作之前,先对它们进行高亮显示。

```python
from selenium import webdriver
import unittest


class ExecuteJavaScriptTest (unittest.TestCase):
    def setUp(self):
        # create a new Firefox session
        self.driver = webdriver.Firefox()
        self.driver.implicitly_wait(30)
        self.driver.maximize_window()

        # navigate to the application home page
        self.driver.get("http://demo.magentocommerce.com/")

    def test_search_by_category(self):

        # get the search textbox
        search_field = self.driver.find_element_by_name("q")
        self.highlightElement(search_field)
        search_field.clear()

        # enter search keyword and submit
        self.highlightElement(search_field)
        search_field.send_keys("phones")
        search_field.submit()
```

```python
        # get all the anchor elements which have product names
        # displayed currently on result page using
        # find_elements_by_xpath method
        products = self.driver.find_elements_by_xpath("//h2[@ class='product-name']/a")

        # check count of products shown in results
        self.assertEqual(2, len(products))

    def tearDown(self):
        # close the browser window
        self.driver.quit()


    def highlightElement(self, element):
        self.driver.execute_script("arguments[0].setAttribute('style',
        arguments[1]);",
        element, "color: green;
        border: 2px solid green;")
        self.driver.execute_script("arguments[0].setAttribute('style',
        arguments[1]);",
        element , "")

if __name__ == "__main__":
    unittest.main(verbosity=2)
```

我们可以通过调用 WebDriver 类的 execute_script 方法来执行 JavaScript 代码，也可以通过这个方法传递参数给 JavaScript 代码，示例代码如下。在这个例子中，我们修改了边框样式，然后又立即恢复到原来的样子。在执行期间，元素将以绿色边框高亮显示，这对于了解哪一个步骤正在执行是非常有用的。

```python
def highlightElement(self, element):
    self.driver.execute_script("arguments[0].setAttribute('style',
    arguments[1]);",
    element, "color: green; border: 2px solid green;")
    self.driver.execute_script("arguments[0].setAttribute('style',
    arguments[1]);",
```

```
element , "")
```

9.3 屏幕截图

自动测试执行过程中，在出错时捕获屏幕截图，是我们在跟开发人员探讨错误时的重要依据。WebDriver 内置了一些在测试执行过程中捕获屏幕并保存的方法，如下表所示。

方法	描述	参数	示例
save_screenshot(filename)	获取当前屏幕截图并保存为指定文件	filename：指定保存的路径/图片文件名	Driver.save_screenshot("homepage.png")
get_screenshot_as_base64()	获取当前屏幕截图 base64 编码字符串（用于 HTML 页面直接嵌入 base64 编码图片）		driver.get_screenshot_as_base64()
get_screenshot_as_file(filename)	获取当前的屏幕截图，使用完整的路径。如果有任何 IOError，返回 False，否则返回 True	filename：指定保存的路径/图片文件名	driver.get_screenshot_as_file('/results/screenshots/HomePage.png')
get_screenshot_as_png()	获取当前屏幕截图的二进制文件数据		driver.get_screenshot_as_png()

接下来，我们通过屏幕截图来捕获一个测试执行出错的场景。场景中，我们定位一个本来应该显示在主页的元素。如果测试脚本没有发现对应元素，则立即抛出 NoSuchElementException 异常，同时截取当前浏览器窗口截图，我们可以把它作为 bug 的依据发给开发人员定位问题。

```
from selenium import webdriver
import datetime, time, unittest
from selenium.common.exceptions import NoSuchElementException


class ScreenShotTest(unittest.TestCase):
    def setUp(self):
        self.driver = webdriver.Firefox()
        self.driver.get("http://demo.magentocommerce.com/")

    def test_screen_shot(self):
        driver = self.driver
```

```
        try:
            promo_banner_elem = driver.find_element_by_id("promo_ banner")
            self.assertEqual("Promotions", promo_banner_elem.text)
        except NoSuchElementException:
            st = datetime.datetime\
                .fromtimestamp(time.time()).strftime('%Y%m%d_%H%M%S')
            file_name = "main_page_missing_banner" + st + ".png"
            driver.save_screenshot(file_name)
            raise

    def tearDown(self):
        self.driver.close()

if __name__ == "__main__":
    unittest.main(verbosity=2)
```

在上述代码中，当测试脚本找不到"promo_banner"元素时，程序就调用 save_screenshot() 方法来自动截屏，并以我们定义的图片文件名保存在指定的路径下。

```
try:
    promo_banner_elem = driver.find_element_by_id("promo_banner")
    self.assertEqual("Promotions", promo_banner_elem.text)
except NoSuchElementException:
    st = datetime.datetime.fromtimestamp(time.time()). strftime('%Y%m%d_%H%M%S')
    file_name = "main_page_missing_banner" + st + ".png"
    driver.save_screenshot(file_name)
    raise
```

 当我们使用上述截屏方法时，推荐使用包含唯一标识（例如时间戳）的名称，并且保存为 PNG 图片等高压缩图片格式，来控制图片的大小。

9.4 屏幕录制

类似屏幕截图，屏幕录制能够更好地帮助我们记录测试过程中到底发生了什么。录像材料可以作为提交问题时的依据发送给项目相关人员，也可以作为产品的功能演示。

然而，Selenium WebDriver 没有内置录制的功能，所以要依赖 Python 类库中名为 Castro 的工具。这是由 Selenium 创始人 Jason Huggin 设计的。Castro 是基于跨平台屏幕录制工具 Pyvnc2swf 开发的。它使用 VNC 协议录制屏幕并生成 SWF 视频文件。

由于符合 VNC 协议，所以我们还可以实现对远程机器（预装 VNC 相关程序包）的屏幕录制。先安装 PyGame，然后安装 Castro，pip 命令如下。

```
pip install Castro
```

如果 Server 和 Viewer 端都是 Windows 的环境，我们可以选择安装 **TightVNC** 工具。

如果在 Ubuntu 操作系统上，可以依次操作 **Settings | Preference | Remote Desktop**，然后选中 **Allow other users to view your desktop** 复选框。在 Mac 上，我们可以安装 Vine VNC Server 或者在 **System Preferences** 中打开 **Remote Desktop**。

结合之前章节我们设计过的测试脚本，添加屏幕录制功能，代码如下。

```python
import unittest
from selenium import webdriver
from castro import Castro


class SearchProductTest(unittest.TestCase):
    def setUp(self):
        # create an instance of Castro and provide name for the output
        # file
        self.screenCapture = Castro(filename="testSearchByCategory.swf")
        # start the recording of movie
        self.screenCapture.start()

        # create a new Firefox session
        self.driver = webdriver.Firefox()
        self.driver.implicitly_wait(30)
        self.driver.maximize_window()

        # navigate to the application home page
        self.driver.get("http://demo.magentocommerce.com/")
```

```python
    def test_search_by_category(self):

        # get the search textbox
        search_field = self.driver.find_element_by_name("q")
        search_field.clear()

        # enter search keyword and submit
        search_field.send_keys("phones")
        search_field.submit()

        # get all the anchor elements which have product names
        # displayed
        # currently on result page using find_elements_by_xpath method
        products = self.driver.find_elements_by_xpath("//h2[@class='product-name']/a")

        # check count of products shown in results
        self.assertEqual(2, len(products))

    def tearDown(self):
        # close the browser window
        self.driver.quit()
        # Stop the recording
        self.screenCapture.stop()

if __name__ == '__main__':
    unittest.main(verbosity=2)
```

从代码中可以看到，要创建一个录制屏幕的会话，我们需要创建一个 Castro 对象并且使用录像文件的路径和名称作为参数初始化实例。start()和 stop()方法用于控制屏幕录制的起止位。代码中 setUp()方法的部分就是一个最佳的初始化 Castro 实例，并且是开始录制的示例。

```python
    def setUp(self):
        # Create an instance of Castro and provide name for the output
        # file
```

```
self.screenCapture = Castro(filename="testSearchByCategory.swf")
# Start the recording of movie
self.screenCapture.start()

# create a new Firefox session
self.driver = webdriver.Firefox()
self.driver.implicitly_wait(30)
self.driver.maximize_window()

# navigate to the application home page
self.driver.get("http://demo.magentocommerce.com/")
```

在 teadDown()部分，我们可以看到当完整的测试用例都执行完成后，调用 stop()方法来停止屏幕录制。代码如下：

```
def teadDown(self):
    # close the browser window
    self.driver.quit()
    # Stop the recording
    self.screenCapture.stop()
```

特别是在组合多个测试场景的测试类中，我们同样也可用上述 setUp() 和 tearDown()方法，来实现整个测试类的屏幕录制操作的开启与停止，无须对不同测试场景重复单独构建。

9.5 弹出窗的处理

弹出窗的处理过程包括：通过弹出窗的名称或句柄来定位，切换 Driver Context 至所需的弹出窗，在弹出窗上执行相关操作步骤，最后跳转回到上级窗口（页面）。

结合我们的测试，创建一个基于浏览器的实例，基于父窗口随后弹出新的窗口，我们统称为子窗口或弹出窗。只要该弹出窗属于当前 WebDriver Context，我们都可以对它进行操作。

下图展示一个弹出窗的例子。

第 9 章 Selenium WebDriver 的高级特性

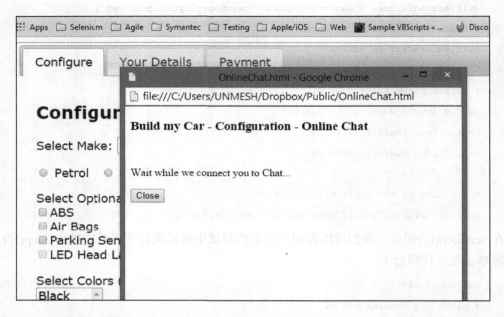

创建一个新的测试类 `PopupWindowTest`，其中包括 `test_popup_window()` 方法，代码如下。

```
from selenium import webdriver
import unittest




class PopupWindowTest(unittest.TestCase):

    URL = "rawgit/upgundecha/learnsewithpython/master/pages/Config.html"

    def setUp(self)        :
        self.driver = webdriver.Firefox()
        self.driver.get(self.URL)
        self.driver.maximize_window()

    def test_popup_ window(self):
        driver = self.driver
```

```python
        # save the WindowHandle of Parent Browser Window
        parent_window_id = driver.current_window_handle

        # clicking Help Button will open Help Page in a new Popup
        # Browser Window
        help_button = driver.find_element_by_id("helpbutton")
        help_button.click()
        driver.switch_to.window("HelpWindow")
        driver.close()
        driver.switch_to.window(parent_window_id)

    def tearDown(self):
        self.driver.close()

if __name__ == "__main__":
    unittest.main(verbosity=2)
```

在 Context 调用弹出窗口显示之前，我们先通过 current_window_handle 属性将父窗口的句柄信息保存下来（稍后我们将使用这个信息从弹出窗返回到父窗口）。接着使用 WebDriver 下的 switch_to.window() 方法获取弹出窗的名称或句柄信息，切换到我们要操作的那个弹出窗（子窗口）。下面我们演示通过名称定位弹出窗。

```
driver.switch_to_window("HelpWindow")
```

我们操作完 Help 窗口之后，通过 close() 方法关闭窗口，并且返回至父窗口，代码如下。

```
driver.close()

# switch back to Home page window using the handle
driver.switch_to_window(default_window)
```

9.6 操作 cookies

为了更好的用户体验，cookies 作为 Web 应用一项很重要的手段，将一些诸如用户偏好、登录信息以及各种客户端细节信息，记录并保存在用户计算机本地。WebDriver 提供了一组操作 cookies 的方法，包括读取、添加和删除 cookies 信息。这些方法可以帮助我们操作 cookies，来校验 Web 应用程序对应的响应。具体方法见下表。

方法	描述	参数	示例
add_cookie(cookie_dict)	在当前会话中添加 cookie 信息	cookie_dict:字典对象，包含 name 与 value 值	driver.add_cookie({"foo","bar"})
delete_all_cookies()	在当前会话中删除所有 cookie 信息		driver.delete_all_cookies()
delete_cookie(name)	删除单个名为 name 的 cookie 信息	name:要删除的 cookie 的名称	driver.delete_cookie("foo")
get_cookie(name)	返回单个名为 name 的 cookie 信息。如果没有找到，返回 none	name:要查找的 cookie 的名称	driver.get_cookie("foo")
get_cookies()	返回当前会话所有的 cookie 信息		driver.get_cookies()

接下的例子，我们来验证用户在首页选择语言后，是否被正确保存至 cookie 中。

```python
import unittest
from selenium import webdriver
from selenium.webdriver.support.ui import Select


class CookiesTest(unittest.TestCase):
    def setUp(self):
        # create a new Firefox session
        self.driver = webdriver.Firefox()
        self.driver.implicitly_wait(30)
        self.driver.maximize_window()

        # navigate to the application home page
        self.driver.get("http://demo.magentocommerce.com/")

    def test_store_cookie(self):
        driver = self.driver
        # get the Your language dropdown as instance of Select class
```

```python
        select_language = \
            Select(self.driver.find_element_by_id("select-language"))

        # check default selected option is English
        self.assertEqual("ENGLISH", select_language.first_selected_option.text)
        # store cookies should be none
        store_cookie = driver.get_cookie("store")
        self.assertEqual(None, store_cookie)

        # select an option using select_by_visible text
        select_language.select_by_visible_text("French")

        # store cookie should be populated with selected country
        store_cookie = driver.get_cookie("store")['value']
        self.assertEqual("french", store_cookie)

    def tearDown(self):
        # close the browser window
        self.driver.quit()

if __name__ == '__main__':
    unittest.main(verbosity=2)
```

上述代码中，我们传递一个 cookie 的名称，就可以通过 get_cookie() 方法获取到对应 cookie 的值。

9.7 章节回顾

在本章，我们学习了一些关于 Selenium WebDriver 的高级特性，例如键盘和鼠标事件、截屏、（屏幕）录制以及操作 cookies。

用 ActionChains 类模拟各种键盘和鼠标的操作，这在处理大量使用键盘和鼠标操作的应用程序时非常有用。

从测试中，你已经看到了如何运行 JavaScript 代码。这是一个非常强大的功能特性，让我们能够轻松应对 Ajax 应用，能够在测试脚本中调用底层的 JavaScipt API。

当测试过程中产生错误（有可能是测试脚本的问题，也有可能是产品的 bug）时，自动截屏或是屏幕录制，都能极大地帮助我们调试测试脚本以及作为提交 bug 的重要依据。

最后部分，我们还学习了操作浏览器窗口与 cookies 的方法。

在下一章节，我们将学习如何将自动化测试与其他持续集成的工具进行联动与整合。

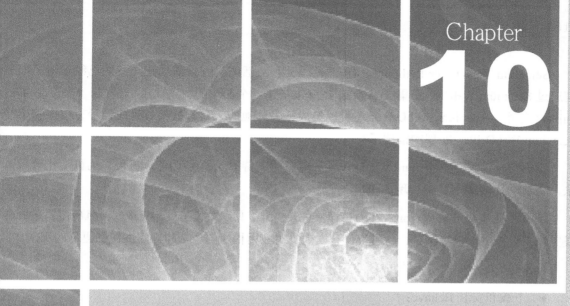

第 10 章
第三方工具与框架集成

Selenium WebDriver Python API 是非常强大和灵活的。到目前为止,我们已经学习了如何通过 Selenium WebDriver 集成 unittest 类库来搭建一个简单的自动化测试框架。然而,除了 unittest 之外,Selenium WebDriver 还可以集成很多其他的工具和框架。目前已经有很多基于 Selenium WebDriver 实现的框架了。

我们可以通过使 Selenium WebDriver 与现有的支持 BDD(行为驱动开发)的框架结合起来,在自动化测试项目中实现 BDD。

还可以将 Selenium Python API 与持续集成(CI)工具、构建工具相集成,一旦应用程序开发完成就可以立即执行测试。这可以使开发人员对应用程序的质量和稳定性得到更早的反馈。

本章包含以下一些主要集成的实例:

- 下载和安装 Behave;
- 使用 Behave 编写 feature;
- 使用 Behave 和 Selenium WebDriver 自动化验证 feature;
- 下载和安装 Jenkins;
- 搭建 Jenkins 运行 Selenium 测试;
- 配置 Jenkins 捕捉测试结果。

10.1 行为驱动开发(BDD)

BDD 是 Dan North 在他的论文《Introducing BDD》中提出的一种敏捷软件开发方法。

BDD 也称为验收测试驱动开发(ATDD)、基于用户故事敏捷测试(story testing)或实例化需求(specification by example)。BDD 鼓励软件项目中的开发者、QA 和非技术人员或商业参与者之间的协作,一起定义项目规范,决定验收标准,用自然语言书写出非程序员可读的测试用例。

Python 中有很多工具都可以实现 BDD,其中两个主要的工具是 Behave 和 Lettuce。
Lettuce 是受到了著名的 Ruby BDD Cucumber 启发。

10.1.1 Behave 安装

安装 Behave 的过程很简单，通过以下命令可直接下载和安装 Behave。

```
pip install behave
```

在命令执行的过程中，会下载并安装 Behave 和它依赖的第三方包。

10.1.2 第一个 feature

在 Behave 中编写第一个 feature 的过程从讨论和列举开发中的应用程序的 feature 和 User Story 开始。各利益相关者聚在一起，有开发者、测试人员、需求分析师和客户，使用各参与者都能理解的通用语言创建 feature、User Story 和验收标准的列表。Behave 支持用 Given-When-Then（GWT）格式的 Gherkin 语言创建 feature 文件。

让我们从示例应用程序中的搜索功能的 feature 开始。该搜索 feature 是让用户从主页搜索产品。 feature 文件以 GWT 格式对 User Story 和验收标准进行简单描述，作为一个场景大纲（Scenario Outline），也称为场景步骤（Scenario Steps），解释如下。

- **Given**：设置一个场景执行的前提条件，本例中——导航到主页。
- **When**：包含一个场景所要执行的操作，本例中——搜索某件产品。
- **Then**：包含一个场景执行后的结果，本例中——检查所有匹配的产品列表能否正常显示。

一个场景中可以有多个 When 和 Then。

```
Feature: I want to search for products

  Scenario Outline: Search
    Given I am on home page
    when I search for "phone"
    then I should see list of matching products in search results
```

我们需要将它保存为扩展名为".feature"的纯文本文件才能在 Behave 中使用。现在新建一个 bdd/feature 文件夹，然后将 feature 文件命名为"search.feature"，保存在该文件夹下。

10.1.2.1 step 定义

feature 文件完成后，我们需要为 feature 文件的场景大纲中的 step 分别进行定义。一个 step 定义就是一个 Python 代码块，代码块用简单明了的文字命名，代码调用 Python API 或者 Selenium WebDriver 命令来执行该 step 的内容。step 定义文件须保存在 feature 文件所在路径的子目录"steps"下。下面就创建一个 search_steps.py 文件来定义上面 feature 文件中的 step。

```python
from behave import *

@given('I am on home page')
def step_i_am_on_home_page(context):
    context.driver.get("http://demo.magentocommerce.com/")


@when('I search for {text}')
def step_i_search_for(context, text):
    search_field = context.driver.find_element_by_name("q")
    search_field.clear()

    # enter search keyword and submit
    search_field.send_keys(text)
    search_field.submit()


@then('I should see list of matching products in search results')
def step_i_should_see_list(context):
    products = context.driver.\
        find_elements_by_xpath("//h2[@class='product-name']/a")
    # check count of products shown in results
    assert len(products) > 0
```

对于每个 GWT，我们都需要创建一个匹配的 step 定义。下面的代码是为 Given 中的"I am on home page"创建的 step 定义。使用@修饰符和 feature 文件中给出的 GWT 来标识对应的 step，如@given、@when、@then，并且接收一个字符串，该字符串包含对应场景 step 中剩余部分的步骤描述信息，如本例中的"I am on home page"。

```python
@given('I am on home page')
def step_i_am_on_home_page(context):
```

```
context.driver.get("http://demo.magentocommerce.com/")
```

我们也可以将 step 中内嵌的参数传递给 step 定义。例如，在@when 中我们这么写上面的搜索语句：when I search for "phone"。然后在 step 定义中通过{text}来读取这个 "phone" 值，如下代码所示。

```
@when('I search for {text}')
def step_i_search_for(context, text):
    search_field = context.driver.find_element_by_name("q")
    search_field.clear()

    # enter search keyword and submit
    search_field.send_keys(text)
    search_field.submit()
```

从上面代码中就可以看到传递给 step 定义的上下文变量。Behave 通过这个上下文变量来存储要共享的信息。它运行在 3 个层面，由 Behave 自动管理。我们还可以使用上下文变量来存储和共享 step 之间的信息。

10.1.2.2 环境配置

在运行 feature 之前，需要创建一个环境配置文件，用于配置 Behave 的常用设置，以及 step 之间或 step 定义文件之间的共享代码。如果用 Selenium WebDriver 和 Firefox 浏览器执行测试步骤的话，这个文件可以在启动 Firefox 时初始化 WebDriver。下面我们在 feature 文件目录中新建一个环境配置文件 environment.py，添加 before_all()和 after_all()方法，这两个方法分别在 feature 执行前和结束后运行。

```
from selenium import webdriver

def before_all(context):
    context.driver = webdriver.Chrome()

def after_all(context):
    context.driver.quit()
```

10.1.2.3 执行 feature

现在可以开始使用 Behave 执行 feature 文件了。操作非常简单，从命令行中进入我们之前创建的 bdd 文件夹目录下，然后执行 "behave" 命令。

behave

Behave 将执行 bdd 文件夹下的所有 feature 文件，通过前面的 step 定义和环境配置信息来运行相应的 scenario。下图是 behave 命令的执行结果。

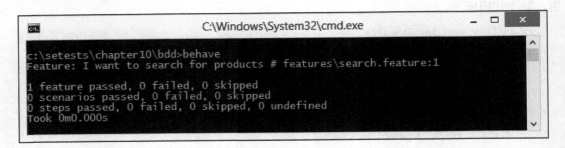

Behave 显示 3 个级别的结果，即 feature、scenario 和 step 各级别通过和失败的数量。

10.1.2.4　使用场景大纲

有时我们可能想用同样的测试步骤、类似数据驱动测试来运行一批已知状态、执行操作和期望结果的测试集合。对于这样的需求，我们可以写出如下的场景大纲。

我们用下面步骤中所给的例子重写 search.feature 文件的场景大纲，这个场景大纲根据 "Example" 部分中的数据，像模板一样工作。

（1）本例创建了两个检验搜索功能的例子：根据"类别"搜索和根据"特定产品名"搜索。"Example"部分以表格形式给出搜索项和预期结果。

```
Feature: I want to search for products

  Scenario Outline: Search
    Given I am on home page
    when I search for <term>
    then I should see results <search_count> in search results


Examples: By category
  |term     | search_count |
  |Phones   | 2            |
  |Bags     | 7            |


Examples: By product name
```

```
| term            | search_count |
| Madison earbuds | 3            |
```

(2) 修改 search_steps.py 文件以匹配上面的 step。

```python
from behave import *

@given('I am on home page')
def step_i_am_on_home_page(context):
    context.driver.get("http://demo.magentocommerce.com/")

@when('I search for {text}')
def step_i_search_for(context, text):
    search_field = context.driver.find_element_by_name("q")
    search_field.clear()

    # enter search keyword and submit
    search_field.send_keys(text)
    search_field.submit()

@then('I should see results {text} in search results')
def step_i_should_see_results(context, text):
    products = context.driver.\
        find_elements_by_xpath("//h2[@class='product-name']/a")
    # check count of products shown in results
    assert len(products) >= int(text)
```

当执行 feature 时，Behave 读取 search.feature 文件中 "Example" 部分数据的行数并自动循环执行场景大纲。它将 "Example" 中的数据传递给 scenario 中的 step，然后执行相应 step 定义中的命令。下图是 Behave 执行修改后的 feature 文件的结果，并打印出了 feature 上运行的所有组合信息。

Behave 也支持用命令行选项 -junit 生成 junit 格式的报告。

```
C:\Users\amitr\Desktop\setests_final\bdd>behave
Feature: I want to search for products # features\search.feature:1

  Scenario Outline: Search                          # features\search.feature:3
    Given I am on home page                         # steps\search_steps.py:4
    When I search for Phones                        # steps\search_steps.py:9
    Then I should see results 2 in search results   # steps\search_steps.py:18

  Scenario Outline: Search                          # features\search.feature:3
    Given I am on home page                         # steps\search_steps.py:4
    When I search for Bags                          # steps\search_steps.py:9
    Then I should see results 7 in search results   # steps\search_steps.py:18

  Scenario Outline: Search                          # features\search.feature:3
    Given I am on home page                         # steps\search_steps.py:4
    When I search for Madison earbuds               # steps\search_steps.py:9
    Then I should see results 3 in search results   # steps\search_steps.py:18

1 feature passed, 0 failed, 0 skipped
3 scenarios passed, 0 failed, 0 skipped
9 steps passed, 0 failed, 0 skipped, 0 undefined
Took 0m14.539s

C:\Users\amitr\Desktop\setests_final\bdd>
```

10.2 持续集成 Jenkins

Jenkins 是 Java 编写的流行的持续集成（CI）服务，起源于 Hudson 项目。Jenkins 和 Hudson 功能相似。

Jenkins 支持各种版本的控制工具，如 CVS、SVN、Git、Mercurial、Perforce 和 ClearCase，而且可以执行用 Apache Ant 或 Java Maven 构建的项目。同时，它也可以利用一些插件、shell 脚本和 Windows 批处理命令来构建其他平台的项目。

Jenkins 除了构建软件功能外，还可以用于搭建自动化测试环境，实现 Selenium WebDriver 测试在无人值守的情况下按照预定的时间调度运行，或每次代码变更提交至版本控制系统时实现自动运行的效果。

在接下来的部分，我们将学习如何搭建 Jenkins 并创建一个自由风格的软件项目来执行测试。

10.2.1 Jenkins 环境准备

为了能够成功使用 Jenkins 执行测试，我们需要做一些修改。我们的目标是在 Jenkins 上按计划时间执行测试，然后收集测试结果并显示在 Jenkins Dashboard 页面。为了实现这个目标，我们将重用第 2 章中创建的冒烟测试。

我们使用了 unittest 的 TestSuite Runner 批量执行测试，并以 JUnit 报告的格式输出测试结果。这就要有 xmlrunner 的 Python 库的支持，可以从 pypi.python 网站获取。

现在执行以下命令下载和安装 xmlrunner。

```
pip install xmlrunner
```

冒烟测试脚本 smoketests.py 是通过 TestSuite Runner 运行 homepagetests.py 和 searchtest.py 两个脚本中的测试的。我们将使用 xmlrunner.XML TestRunner 来运行冒烟测试并生成 JUnit 测试报告。此报告以 XML 格式生成，保存在 test-reports 子文件夹中。要使用 xmlrunner，需要对 smoketest.py 做些改动，如以下代码的**高亮**显示部分。

```
import unittest
from xmlrunner import xmlrunner
from searchtest import SearchProductTest
from homepagetests import HomePageTest


# get all tests from SearchProductTest and HomePageTest class
search_tests = unittest.TestLoader().loadTestsFromTestCase(SearchProd uctTest)
home_page_tests = unittest.TestLoader().loadTestsFromTestCase(HomePag eTest)


# create a test suite combining search_test and home_page_test
smoke_tests = unittest.TestSuite([home_page_tests, search_tests])


# run the suite
xmlrunner.XMLTestRunner(verbosity=2, output='test-reports').run(smoke_ tests)
```

10.2.2 搭建 Jenkins

搭建 Jenkins 相当简单。你可以下载各种平台的 Jenkins 包并进行安装。在下面的例子中，我们将安装并启动 Jenkins，然后创建一个新的构建作业以对示例应用程序进行冒烟测试。

（1）下载并安装 Jenkins CI 服务器。我们下载的是 Jenkins Windows 安装包，并在 Windows 7 上安装 Jenkins。

（2）从浏览器中进入 **Jenkins Dashboard** 页面（默认情况下为 http:// localhost:8080）。

（3）在 **Jenkins Dashboard** 页面上，单击新建项目（**New Item**）或创建新作业（**create new jobs**）链接，创建一个新的 Jenkins 作业，如下图所示。

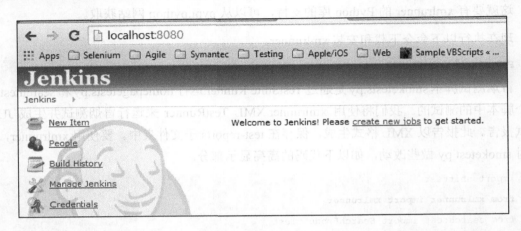

（4）在项名称（**Item name**）文本框中输入 Demo_App_Smoke_Test，然后选择构建自由风格的软件项目（**Freestyle project**）单选按钮，如下图所示。

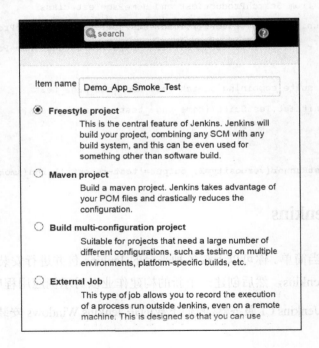

（5）单击确定（OK）按钮。以上面指定名称命名的新作业就创建成功了。

> 我们可以连接至各种版本的控制或源代码控制管理（SCM）工具，如 SVN、Git、Perforce 等，以存储源代码和测试代码。然后作为构建步骤的一部分，获取最新版本的代码，并在 Jenkins 中构建和测试软件。但是，在本示例中，为了保证过程的简洁性，我们将在执行 Windows 批处理命令（Execute Windows batch command）构建步骤中将当前文件夹下的测试脚本复制到 Jenkins 工作空间下，如以下步骤中所述。

（6）在构建（Bulid）部分中，单击添加构建步骤（Add build step），然后从下拉列表中选择执行 Windows 批处理命令（Execute Windows batch command）选项，如下图所示。

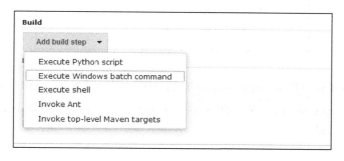

（7）在命令（Command）文本框中输入以下命令，如下图所示。在不同的电脑上路径可能会有所不同。这个命令将冒烟测试的 Python 脚本文件复制到 Jenkins 工作空间下并执行 smoketest.py。

```
copy c:\setests\chapter10\smoketests\*.py
python smoketest.py
```

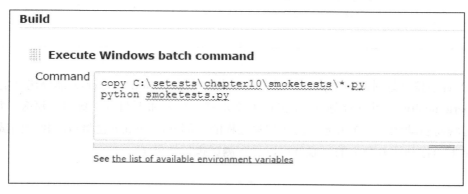

（8）我们在前面已经配置了 smoketest.py 以生成 JUnit 格式的测试结果，并将测试结果显示在 Jenkins Dashboard 页面。要在 Jenkins 中集成这些报告，先单击添加构建后操作（Add post-build action），然后选择发布 JUnit 测试结果报告（Publish JUnit test result report）选项，如下图所示。

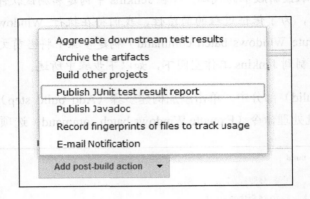

（9）在构建后操作（Post-build Actions）部分中，在测试报告 XML（Test report XMLs）文本框中添加 test-reports/*.xml，如下图所示。Jenkins 每次运行测试的时候，它将从 test-reports 子文件夹中读取测试结果。

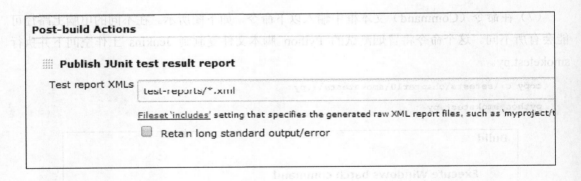

（10）若想按计划时间自动执行测试，在构建触发器（Build Triggers）部分选择定期构建（Build periodically），并在计划（Schedule）文本框中输入如下图所示数据。那么，每天 22 点构建过程将自动触发，作为无人值守构建过程的一部分，Jenkins 也将自动执行测试，这样第二天早上当你到达办公室的时候就可以看到测试执行结果了。

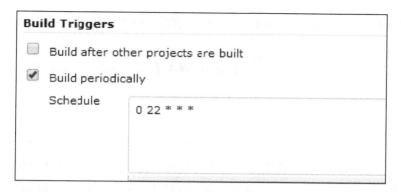

（11）单击保存按钮保存作业配置。Jenkins 将会显示新创建的作业项目页面。

（12）现在可以来检验一下所有的配置项是否设置好，测试是否能成功执行。单击开始构建（Build Now）链接手动运行该作业，如下图所示。

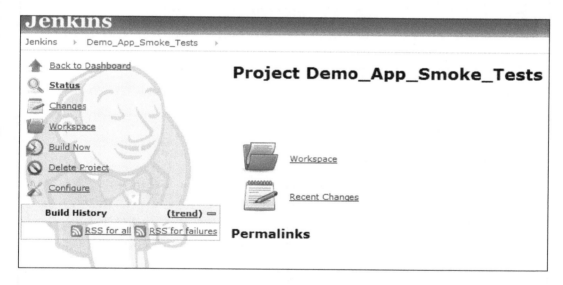

（13）在构建历史（Build History）部分中可以查看构建的运行状态，如下图所示。

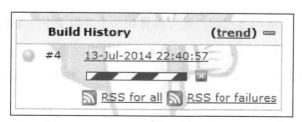

（14）单击构建历史（Build History）部分中正在运行的项目，将打开如下图所示的页面。

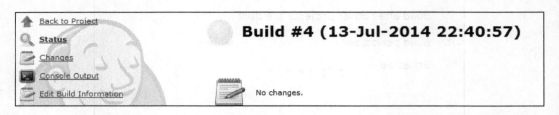

（15）除了 Jenkins 页面上的执行状态和进度条，还可以通过打开控制台输出（Console Output）链接观察后台执行信息。下图是有命令行输出信息的"控制台输出"页面。

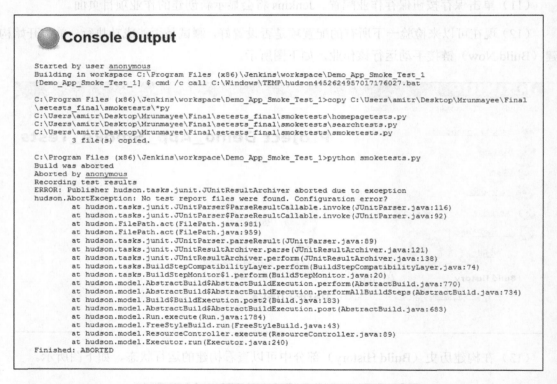

（16）一旦 Jenkins 完成构建过程，就可以看到一个类似于下一个截图所示的构建页面。

（17）Jenkins 通过读取 unittest 框架生成的测试结果文件，在页面上显示测试结果和其他各项指标。单击构建页面上的测试结果（Test Results）链接可以查看 Jenkins 保存的测试结果。

（18）我们之前配置的测试结果以 JUnit 格式生成。当单击测试结果（Test Results）时，

Jenkins 将会显示 JUnit 测试结果，如下图所示，显示测试结果摘要，其中失败的测试会高亮显示。

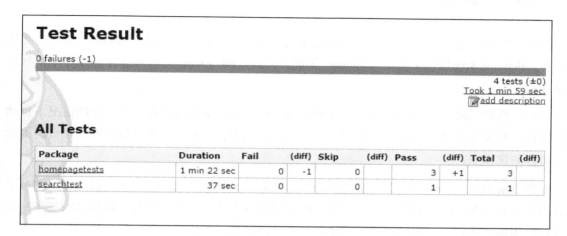

（19）我们也可以单击 Package 名字的链接来查看各个测试的详细结果信息，如下图所示。

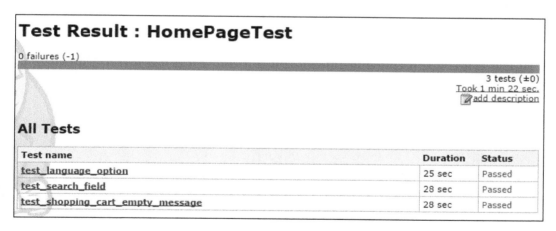

Jenkins 还会以下图所示的格式在 Dashboard 主页上显示所有构建作业的最终状态信息。

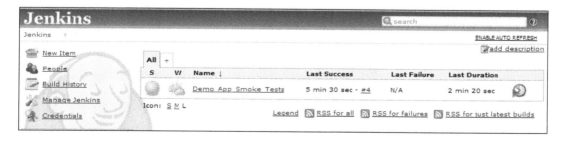

10.3 章节回顾

在本章中，我们学习了如何将 Selenium 与 BDD（Behave）和 CI（Jenkins）集成。实现了将 Selenium WebDriver API 与 Behave 集成，并通过编写 feature 和 step 定义文件来执行自动化验收测试。

通过搭建 Jenkins 运行 Selenium WebDriver 测试，从而实现每晚在无人值守的情况下自动构建程序和执行测试。Jenkins 提供了一个易于搭建的平台，来对接各种程序开发平台的构建和运行测试作业。

到目前为止，我们已经完成了 Python Selenium WebDriver 的学习之旅。我们主要学习了使用 Selenium WebDriver 进行 Web 应用程序在浏览器中的交互测试和自动化测试的基本知识点，运用这些知识点，我们就可以构建自己的自动化测试框架了。

读书笔记

读书笔记